This timely book provides realistic analyses of potential threats to the Olympics and possible countermeasures. All those responsible for protecting the Games and other major public events and gatherings would benefit from reading it.
Paul Wilkinson, Emeritus Professor of International Relations,
St Andrews University

Drawing on the expertise of internationally renowned academics and industry experts, *Terrorism and the Olympics* is an invaluable reference resource for those concerned about preventing potential threats to London's 2012 Olympics and applying best practices to major public event security in general.
Dr Joshua Sinai, Virginia Tech

Terrorism and the Olympics

This book aims to outline the progress, problems and challenges of delivering a safe and secure Olympics in the context of the contemporary serious and enduring terrorist threat. The enormous media profile and symbolic significance of the Olympic Games, the history of terrorists aiming to use such high-profile events to advance their cause, and Al Qaeda's aim to cause mass casualties, all have major implications for the security of London 2012.

Drawing on contributions from leading academics and practitioners in the field, the book will assess the current terrorist threat, particularly focusing on terrorist targeting and how the Olympics might feature in this, before addressing particular response themes such as transport security, the role of surveillance, resilient designing of Olympic sites, the role of private security and the challenge of inter-agency coordination. The book will conclude by providing an assessment of the legacy of Olympic security to date and will discuss the anticipated issues and dilemmas of the future.

This book will be of interest to students of terrorism studies, security studies, counter-terrorism and sports studies.

Anthony Richards is a Senior Lecturer in Terrorism Studies and Programme Leader for the BA (Hons) Criminology and Criminal Justice degree at the University of East London. He has written on a wide variety of terrorism-related themes.

Pete Fussey is a Senior Lecturer in Criminology at the University of Essex and has authored numerous papers and articles on Olympic security.

Andrew Silke is the Field Leader for Criminology and the Director of Terrorism Studies at the University of East London. He is author of several books and over 80 articles on terrorism.

Political Violence
Edited by Paul Wilkinson and David Rapoport

This book series contains sober, thoughtful and authoritative academic accounts of terrorism and political violence. Its aim is to produce a useful taxonomy of terror and violence through comparative and historical analysis in both national and international spheres. Each book discusses origins, organisational dynamics and outcomes of particular forms and expressions of political violence.

Aviation Terrorism and Security
Edited by Paul Wilkinson and Brian M. Jenkins

Counter-Terrorist Law and Emergency Powers in the United Kingdom, 1922–2000
Laura K. Donohue

The Democratic Experience and Political Violence
Edited by David C. Rapoport and Leonard Weinberg

Inside Terrorist Organizations
Edited by David C. Rapoport

The Future of Terrorism
Edited by Max Taylor and John Horgan

The IRA, 1968–2000
An analysis of a secret army
J. Bowyer Bell

Millennial Violence
Past, present and future
Edited by Jeffrey Kaplan

Right-Wing Extremism in the Twenty-First Century
Edited by Peter H. Merkl and Leonard Weinberg

Terrorism Today
Christopher C. Harmon

The Psychology of Terrorism
John Horgan

Research on Terrorism
Trends, achievements and failures
Edited by Andrew Silke

A War of Words
Political violence and public debate in Israel
Gerald Cromer

Root Causes of Suicide Terrorism
Globalization of martyrdom
Edited by Ami Pedahzur

Terrorism versus Democracy
The liberal state response, 2nd edition
Paul Wilkinson

Countering Terrorism and WMD
Creating a global counter-terrorism network
Edited by Peter Katona, Michael Intriligator and John Sullivan

Mapping Terrorism Research
State of the art, gaps and future direction
Edited by Magnus Ranstorp

The Ideological War on Terror
Worldwide strategies for counter-terrorism
Edited by Anne Aldis and Graeme P. Herd

The IRA and Armed Struggle
Rogelio Alonso

Homeland Security in the UK
Future preparedness for terrorist attack since 9/11
Edited by Paul Wilkinson et al.

Terrorism Today, 2nd Edition
Christopher C. Harmon

Understanding Terrorism and Political Violence
The life cycle of birth, growth, transformation, and demise
Dipak K. Gupta

Global Jihadism
Theory and practice
Jarret M. Brachman

Combating Terrorism in Northern Ireland
Edited by James Dingley

Leaving Terrorism Behind
Individual and collective disengagement
Edited by Tore Bjørgo and John Horgan

Unconventional Weapons and International Terrorism
Threat convergence in the twenty-first century
Edited by Magnus Ranstorp and Magnus Normark

International Aviation and Terrorism
Evolving threats, evolving security
John Harrison

Walking Away from Terrorism
John Horgan

Understanding Violent Radicalisation
Terrorist and jihadist movements in Europe
Edited by Magnus Ranstorp

Terrorist Groups and the New Tribalism
Terrorism's Fifth Wave
Jeffrey Kaplan

Negotiating with Terrorists
Strategy, tactics and politics
Edited by I. William Zartman and Guy Olivier Faure

Explaining Terrorism
Causes, processes and consequences
Martha Crenshaw

The Psychology of Counter-Terrorism
Edited by Andrew Silke

Terrorism and the Olympics
Major event security and lessons for the future
Edited by Anthony Richards, Pete Fussey and Andrew Silke

Terrorism and the Olympics

Major event security and lessons for the future

Edited by Anthony Richards,
Pete Fussey and Andrew Silke

LONDON AND NEW YORK

First published 2011
by Routledge
2 Park Square, Milton Park, Abingdon, Oxon OX14 4RN

Simultaneously published in the USA and Canada
by Routledge
711 Third Avenue, New York, NY 10017

Routledge is an imprint of the Taylor & Francis Group, an informa business

First issued in paperback 2012

© 2011 Anthony Richards, Pete Fussey and Andrew Silke for selection and editorial matter; individual contributors, their contributions

The right of the editors to be identified as the authors of the editorial material, and of the authors for their individual chapters, has been asserted in accordance with sections 77 and 78 of the Copyright, Designs and Patents Act 1988.

Typeset in Times by Wearset Ltd, Boldon, Tyne and Wear

All rights reserved. No part of this book may be reprinted or reproduced or utilised in any form or by any electronic, mechanical, or other means, now known or hereafter invented, including photocopying and recording, or in any information storage or retrieval system, without permission in writing from the publishers.

British Library Cataloguing in Publication Data
A catalogue record for this book is available from the British Library

Library of Congress Cataloging-in-Publication Data
A catalog record has been requested for this book

ISBN13: 978-0-415-49939-2 (hbk)
ISBN13: 978-0-203-83522-7 (ebk)
ISBN13: 978-0-415-53242-6 (pbk)

Contents

Notes on contributors ix

1 **Towards an understanding of terrorism and the Olympics** 1
 ANTHONY RICHARDS, PETE FUSSEY AND ANDREW SILKE

PART I
The terrorist threat 13

2 **Terrorism, the Olympics and sport: recent events and concerns for the future** 15
 ANTHONY RICHARDS

3 **Al Qaeda and the London Olympics** 32
 AFZAL ASHRAF

4 **Understanding terrorist target selection** 49
 ANDREW SILKE

PART II
Response themes: transport security, the role of surveillance and designing stadia for safer events 73

5 **Securing the transport system** 75
 STEVE SWAIN

6 **Surveillance and the Olympic spectacle** 91
 PETE FUSSEY

7 **Strategic security planning and the resilient design of Olympic sites** 118
JON COAFFEE

PART III
Coordination, roles and responsibilities 133

8 **Governing the Games in an age of uncertainty: the Olympics and organisational responses to risk** 135
WILL JENNINGS

9 **The role of the private security industry** 163
DAVID EVANS

10 **The challenge of inter-agency coordination** 180
KEITH WESTON

11 **The European Union and the promotion of major event security within the EU area** 208
FRANK GREGORY

12 **Critical reflections on securing the Olympics: conclusions and ways forward** 227
PETE FUSSEY, ANTHONY RICHARDS AND ANDREW SILKE

13 **Terrorist threats to the Olympics, 1972–2016** 239
PETE FUSSEY

Index 244

Contributors

Editors

Pete Fussey is a Senior Lecturer in Criminology at the University of Essex, UK, and lectures on a range of criminological themes – including criminological theory, transnational organised crime, urbanisation and crime and psychological criminology – and on critical terrorism and counter-terrorism related issues. Dr Fussey's main research interest focuses on technological surveillance, particularly in relation to their deployment in relation to crime and terrorism, and has published widely on the subject. Currently, he is researching the form and impact of the 2012 Olympic security strategy and, also, conducting ethnographic research into organised crime and the informal economy in East London. He has also researched terrorism and counter-terrorism for a number of years. At present he is working on two large-scale funded research projects: examining counter-terrorism in the UK's crowded spaces (EPSRC EP/HO230LX/1, £1.2m) and researching the future resilience of the UK's critical infrastructures until 2050 (ESRC & EPSRC EP/I005943/1, £1.7m). Dr Fussey also has two additional books due for publication in 2011: *Securing the Olympic City: Reconfiguring London for 2012 and Beyond*, Ashgate (with J. Coaffee, G. Armstrong and D. Hobbs) and *Researching Crime: Approaches, Method and Application*, Palgrave (with C. Crowther-Dowey).

Anthony Richards is a Senior Lecturer in Terrorism Studies in the School of Law, University of East London, where he teaches on the Critical Perspectives on Terrorism, Critical Perspectives on Counter-Terrorism and Transnational Organised Crime modules. Prior to joining UEL, he was a Senior Research Associate at the Centre for the Study of Terrorism and Political Violence, University of St Andrews, where he taught on the terrorism studies distance learning programme for which he designed two modules: Terrorist Ideologies, Aims and Motivations; and Terrorist Modus Operandi. He also worked on the UK Economic and Social Research Council project 'The Domestic Management of Terrorist Attacks in the UK', which was an assessment of both the UK's ability to pre-empt a major terrorist attack and its capacity to deal with the consequences of one (three of his chapters have been published in the book version of the report: *Homeland Security in the UK:*

Future Preparedness for Terrorist Attack Since 9/11, Routledge, 2007). He has also written on British public and Muslim attitudes towards terrorism and UK counter-terrorism ('Countering the Psychological Impact of Terrorism: Challenges for UK Homeland Security', in A. Silke (ed.) *Psychology, Terrorism and Counterterrorism*, London: Routledge, forthcoming 2011) and is currently working on a terrorism textbook (co-authored with Dr Peter Lehr, University of St Andrews) and various articles within the field of terrorism studies. He was Assistant Editor of the academic journal *Terrorism and Political Violence* from 2002–2005.

Andrew Silke is internationally recognised as a leading expert on terrorism in general and terrorist psychology in particular. He has a background in forensic psychology and criminology, and has worked both in academia and for government. His research and writings are published extensively in academic journals, books and the popular press, and he is the author of over 100 publications on subjects relating to terrorism and counter-terrorism. He has worked with a variety of government departments and law enforcement and security agencies. In the United Kingdom these include, the Home Office, the Ministry of Justice, the Ministry of Defence, the UK prison service and several UK police forces. Overseas, he has worked with the United Nations, the US Department of Justice, the US Department of Homeland Security, NATO, the European Defence Agency, the European Commission and the Federal Bureau of Investigation. Professor Silke serves by invitation on the United Nations Roster of Terrorism Experts and the European Commission's European Network of Experts on Radicalisation. He has provided invited briefings on terrorism-related issues to Select Committees of the House of Commons and was appointed in 2009 as a Specialist Advisor to the House of Commons Communities and Local Government Committee for its inquiry into the government's programme for preventing violent extremism. He currently holds a Chair in Criminology at the University of East London where he is the Field Leader for Criminology, and the Programme Director for Terrorism Studies.

Contributors

Afzal Ashraf is a Research Fellow in the Centre for the Study of Terrorism and Political Violence at St Andrews University where his topic of research is Al Qaeda's ideology. He is also Vice Chairman of the Centre's Board of Advisors. He initially began researching Al Qaeda in 1998 as part of a dissertation for a Masters degree in Defence Studies at Kings College London. In 2003 he joined the St Andrews University Centre for the Study of Terrorism and Political Violence where his research resumed into Al Qaeda as the topic of his PhD Thesis. Ashraf has published book chapters and academic journal articles on a range of topics including: The Internet and the Iraqi Insurgency, Security Sector Reform of the Police in Iraq and The Ethics of Bombing and the War on Terror. He is regularly invited to lecture to senior practitioners in Europe, the USA and the Middle East and present papers on contemporary security issues including: Religious Extremism and Terrorism, Culture and Intelligence in Counter Insurgency, and Strategy in Contemporary Conflicts.

Jon Coaffee holds a Chair in Spatial Planning and Urban Resilience in the Centre for Urban Policy Studies, Birmingham Business School, University of Birmingham, UK. Before this appointment he worked at the Universities of Newcastle and Manchester. His work has focused on the interplay of physical and socio-political aspects of resilience. In particular, it has analysed the ability of businesses, governments and communities to anticipate shocks, and ultimately embed resilience within everyday activities and the ability to attain a culture of 'resilience'. This work has been funded by a variety of UK Research Council programmes for 'New Security Challenges' and 'Global Uncertainties'. He has published widely on the social and economic future of cities, and especially the impact of terrorism and other security concerns on the functioning of urban areas. This work has been published in multiple disciplinary areas such as geography, town planning, political science, sports studies and civil engineering. Most notably, he published *Terrorism Risk and the City* (Ashgate, 2003), *The Everyday Resilience of the City: How Cities Respond to Terrorism and Disaster* (Palgrave, 2008) and *Terrorism Risk and the Global City: Towards Urban Resilience* (Ashgate, 2009).

David Evans is a Project Director at the British Security Industry Association (BSIA) where he is responsible for ensuring that the private security industry is a partner with the public sector in the provision of security at major international events. He joined the BSIA in 2006 from Legion Group plc, a national security company, where he was Managing Director. Prior to this, he held positions as Sales and Marketing Director with Securicor Guarding and was Commercial Director for Sterling Granada. His experience in those companies has ranged from security guarding to systems, event security and consultancy. He is a past Chair of the Home Office working group on Safe Deposit security. David joined the security industry in 1981 following service in the Army Intelligence Corps and a period in Special Forces. He is a Freeman of the Worshipful Company of Security Professionals and a Fellow of the Security Institute.

Frank Gregory was Professor of European Security and Jean Monnet Chair in European Political Integration in the Division of Politics and International Relations in the School of Social Sciences at Southampton University until he retired in 2010. His research interests are linked to the homeland security, terrorism, crime and policing aspects of the EU's internal security policy area with special reference to UK-related matters. He has worked on a Falcone Project and two EU-SEC projects. He was co-opted as a member of a specialist ACPO sub-committee (Explosives Detection), he was a member of a UK government advisory panel on emergency responses (SAPER) and he is a Visiting Professor at Cranfield University and a member of RUSI.

Will Jennings is ESRC/Hallsworth Research Fellow at the University of Manchester, and a Research Associate at the ESRC Centre for Analysis of Risk and Regulation at the London School of Economics. His research explores the politics and management of risk in mega-projects and mega-events such as the Olympic Games. His research is currently funded through an ESRC Research Fellowship ('Going for Gold: The Olympics, Risk and Risk Management'). Other research interests include public policy and public opinion, political behaviour, bureaucracy, agenda-setting and blame management by public officeholders. He is also Co-Director of the UK Policy Agendas Project (www.policyagendas.org.uk) which analyses the policy agenda of British governments from 1911 to the present.

Steve Swain is the CEO of the Security Innovation and Technology Consortium (SITC), a role he started in June 2008. Prior to this, he was a consultant with Control Risks, an international risk consultancy. He joined them in September 2006 after retiring as a Chief Superintendent in the MPS. His last post was the Head of the Police International Counter Terrorist Unit (PICTU), a national police and MI5 unit, with responsibility for designing counter-terrorist policing options for the UK. He worked with MI5, Special Branch and the Anti-Terrorist Branch to produce assessments of the national intelligence picture. Steve is a leading authority on suicide terrorism and the architect of the UK tactics to counter the threat from international and domestic terror groups. He was part of the UK team working with the Greek Authorities on the security of the Athens Olympics and he spent time in Beijing performing a similar function for the 2008 Olympics. During his police career, he worked at Heathrow Airport where he had responsibility for the airport's counter-terrorist policing and he worked with the BAA to review and develop new counter-terrorism responses for the airport. He also worked with Transec on detection technologies for mass transport systems.

Keith Weston has, since October 2005, been a Senior Research Fellow in Counter Terrorism within the Security Studies Institute at Cranfield University, located within the UK Defence Academy. He currently lectures on several MSc courses, including the MSc in Resilience, the MSc in Global Security, the MSc in Defence Leadership, the MSc in Forensic Engineering Science as well as on the MBA (Defence). In March 2010 he was an Adjunct Professor on the Programme for Terrorism and Security Studies (PTSS) at the George C. Marshall Centre, Garmisch-Partenkirchen, Germany. After 32 years service, he retired from the Metropolitan Police Service in 2005, with the rank of Detective Chief Superintendent. He served in both the Special Branch and the Anti-Terrorist Branch. Whilst in the Anti-Terrorist Branch, he was involved in the development of the UK multi-agency response to the threat from CBRN. He was involved in the investigation of suspected CBRN and conventional terrorist incidents, including the murder of the UK Defence Attaché in Athens in 2000. His last post was the head of the Police International Counter Terrorism Unit. He is a member of the International Board of Advisors of the New York University Law School and is a non-executive director of the Security Innovation Technology Consortium (SITC). He is a Founding Fellow of the Institute of Civil Protection and Emergency Management and is on the Board of Advisors for CSARN (City Security and Resilience Networks). In 1992 he was awarded an MA in Police Studies by the University of Exeter. In 2003 he was awarded the Queen's Police Medal for distinguished service.

1 Towards an understanding of terrorism and the Olympics

Anthony Richards, Pete Fussey and Andrew Silke

Introduction

One of the enduring principles of the Olympic Games is the promotion of peace through sport. It is deeply disheartening then that the Games themselves have repeatedly been both the location and target of threats and acts of violence in recent decades. Indeed, some of the most high-profile terrorist attacks of the past 50 years have been deliberately targeted against the Olympics, and on the eve of the 2012 Games, there is little sense that such threats to future Olympiads are likely to be rare.

The two most significant terrorist attacks against the Olympics to date remain the Black September attack against the Israeli team at the 1972 Munich Games and Eric Rudolph's nail-bomb attack carried out against the Atlanta Games in 1996. Munich, in particular, remains the benchmark by which all other terrorist attacks against the Olympics are judged. Indeed, for terrorist groups themselves, Munich remains an iconic event. Captured Al Qaeda documents, for example, show that the group regards Munich as the second most important terrorist attack of the past 50 years (not surprisingly, they rate 9/11 as the most important).

Perhaps one of the most surprising features of the Munich attack, though, was just how late planning for the assault started. Black September only decided to try to target the Games at a meeting in Rome on 15 July 1972, barely six weeks before the Opening Ceremony (Wolff, 2002). While this was relatively late in the day to start planning for what would ultimately become such an extraordinary attack, Black September nonetheless still had some reasons to be optimistic. As a group, the organisation had considerable resources to draw upon. Israeli intelligence later judged that at least 40 people were involved in the planning, preparation and execution of the attack. Black September also had the benefit of experience and had carried out a number of other successful terrorist attacks in Europe in the run-up to Munich, including three in West Germany earlier that year.

The terrorist group was also greatly helped by the fact that security surrounding the Munich Games was desperately lax. Indeed, Munich is now widely regarded within security circles as a superb example of what not to do on almost every level when it comes to protecting major events such as the Olympics.

Black September had little difficulty taking advantage of this as they prepared for the attack. Indeed, in the weeks prior to the assault, a number of the terrorists were even able to get jobs working at the Olympic village, which proved incredibly useful as they planned the attack (Reeve, 2001), while others appeared able to gain entry to Olympic sites almost at will. The attack itself was launched during the second week of the Games. On 5 September 1972, the eight-man terrorist team scaled the outer perimeter by pretending to be athletes returning from a party (and were assisted in doing so by genuine athletes who were also sneaking back in). The terrorists then stormed the accommodation area housing the Israeli team, killing two team members and taking a further nine hostage. Negotiations started between the terrorists and the West German authorities but the latter were poorly prepared for the crisis. Although at least one German security planner had specifically raised a terrorist hostage event as one of the risks facing the Games, this had been dismissed by senior figures as far-fetched and no serious precautions or preparations were made (Wolff, 2002).

After several hours of negotiations, the authorities agreed to provide a passenger jet to fly the Black September team and their hostages out of the country, after which an attempt was made by the German police to overpower the terrorists at the airport and free the hostages. Planning and preparation for the rescue attempt, however, was shockingly poor. The police at the airport, for example, had originally been told that there were just five terrorists and based their plans for the rescue on that number. They were only informed that there were actually eight terrorists minutes before the helicopters carrying the terrorists and hostages arrived, by which time it was too late to bring in extra resources or change the plan made. As a result, there were only five police snipers in position at the airport. The snipers themselves were inexperienced and had not received the specialist tactical training normally given to police SWAT officers (the SWAT team that eventually did arrive was sent to the wrong part of the airport and played no part in the crisis).

The resulting shootout, not surprisingly, was a debacle. Over a confused three-hour period, five terrorists, one police officer and all nine hostages were killed at the airport. The three surviving terrorists were captured but were quickly freed when, just three months later, Black September hijacked a Lufthansa jet in order to force the West German government to release them. Still shaken by the Munich experience, the government capitulated swiftly to the terrorists' demands and the three were released (Silke, 2001a).

Israel, however, did not let matters rest there. Outraged by the deaths of their team members and later, no doubt, goaded even further by the release of the three surviving terrorists, the Israeli government secretly authorised Operation Wrath of God. This was a deliberate and systematic campaign of assassination targeting all the individuals believed to be involved in the planning and carrying out of the Munich attack. Wrath of God was not about capturing or imprisoning those responsible. It was purely and simply about killing those the Israelis could find and terrorising those they could not. In order to accomplish this task, a specialist assassination unit – known as the *kidon* – was activated, comprising of just under 40 highly trained members (Silke, 2003).

Death warrants were issued for 35 people believed to have been involved or connected to the Munich attack, and the *kidon* members were allowed enormous latitude in their operations. Less than five weeks after Munich, the first of the targets, Wael Zaitter, a cousin of Yasser Arafat and a principle organiser of PLO terrorism in Europe, was gunned down at his apartment near Rome. Over the following months, the assassination squads killed several more people on the list, and in a major operation on 10 April 1973, a large commando raid was launched against Palestinians based in Beirut. Over 100 people were killed in the attack, including a number on the Wrath of God death list. In time, two of the three terrorists released by West Germany were tracked down and killed, though one managed to survive and is believed to still be in hiding (Reeve, 2001).

Despite the mistaken assassination of at least one innocent victim in Norway – and a growing perception that many on the death list probably had no involvement with Munich – Wrath of God continued until 1979, when assassins finally managed to kill the elusive Ali Hassan Salameh with a car bomb in Beirut. A total of 100,000 people attended his funeral, while the widow of one of the murdered Munich athletes publicly thanked the assassins for what they had done. Although Wrath of God itself was wound down, targeted assassination as a policy remained a key element in Israeli counter-terrorism efforts in the following decades (with questionable impacts).

While Munich is an example of what a large, well-equipped and experienced terrorist group can do when it focuses on the Olympics, the next most significant attack, Atlanta, is a warning that isolated loners can be almost as dangerous. At 1.20 a.m. on Saturday 27 July 1996, a 40 lb bomb exploded in Centennial Olympic Park in Atlanta during the Summer Olympics. The explosion killed one person and injured over 100. A second person died at the scene shortly afterwards due to a heart attack. The bomb had been placed in a backpack and was a home-made device packed with a large number of masonry nails.

Initial suspicion as to who was responsible for the attack fell on the unfortunate Richard Jewell, a security guard who was working in the park and was the first to notice the suspicious backpack. He alerted police to the device and was helping to evacuate the area when the bomb exploded. Initially, Jewell was hailed by the media as a hero, but within days it emerged that the FBI considered the 34-year-old security guard to be the main suspect for the bombing.

Jewell apparently closely matched a psychological profile that described the bomber as a former policeman who longed for heroism. Items taken from Jewell's home during the service of search warrants, such as a collection of newspaper clippings describing Jewell as a hero, added fuel to the media fire. However, the search warrants failed to produce any physical evidence whatsoever to link Jewell to the bombing and the FBI lied to Jewell when they brought him in for questioning. After a prolonged period, Jewell was finally informed that 'barring any newly discovered evidence' he was no longer considered a suspect (Silke, 2001b).

Eventually, the real perpetrator was identified when the Atlanta bombing was linked to a number of subsequent explosions, including two others in Atlanta and a third attack on a women's health clinic in Birmingham. These latter attacks were believed to have been carried out by elements of the extremist 'Army of God' – a Christian fundamentalist fringe group. Particular attention was focused on Eric Rudolph, a key figure within this movement. Rudolph's truck had been seen by witnesses at the scene of the Birmingham bombing, and shortly afterwards a man wearing a wig was seen driving off at speed. The truck was found abandoned and Rudolph disappeared. Forensic examination revealed close similarities in the construction of the explosive devices in all four attacks. After his eventual capture in 2003, Rudolph admitted responsibility for the Atlanta attack, and his reasons for targeting the Olympics are described in more detail later in this volume.

Attacks such as Munich and Atlanta cast long shadows over any effort to understand the Olympics, the threats it faces and the precautions that must be taken. It is no revelation to suggest that the Olympics provides terrorists with a unique opportunity to publicise their goals. Terrorism is a form of communication intended to send a message to target audiences. What better vehicle for global dissemination than the Olympics, whose watching audience are said to amount to billions.? It is estimated that over one billion people watched the opening ceremony for the Beijing Games in 2008, and the total number of hours spent viewing the Summer Games now consistently exceeds over 34 billion for each Games (International Olympic Committee, 2010). Such overwhelming media interest has not been lost on terrorist organisations, with the Munich attack itself representing a telling example. In one day, proclaimed Yasser Arafat, the Palestinian cause gained more attention than any number of international conferences. Munich certainly served as a warning that future Olympic events could be similarly exploited by terrorists for their publicity value if nothing else.

Terrorism and the Olympics today

So why, over three-and-a-half decades later, produce a book on 'Terrorism and the Olympics'? The first and perhaps most significant reason is that the nature of the contemporary terrorist threat has changed. The predominance of nationalist/ separatist and left-wing organisations on the terrorist landscape in the 1970s was being challenged by the emergence of religious terrorism in the 1980s and 1990s. Yet, although religiously driven, organisations like Hezbollah and Hamas have had very real domestic and political goals that arguably placed some restriction on the level of violence that they were willing to perpetrate. As devastating as their use of suicide attacks was, their goal was not to cause massive casualties on the scale of some of the attacks perpetrated by Al Qaeda. Yet, potentially more deadly than Hamas and Hezbollah was the emergence of the Japanese religious cult Aum Shinrikyo. Its sarin gas attack on the Tokyo subway in March 1995 represented a watershed for many emergency services across the world who considered how they would prepare for such an attack. The events of 11 September 2001 also showed what terrorists motivated by religion could be capable of. Brian Jenkins once reasonably stated that 'terrorists want a lot of people watching and a lot of people listening and not a lot of people dead' (Jenkins, cited in Hoffman, 2001). The difference today, however, is that contemporary international terrorism, at least in the form of Al Qaeda and its affiliates, also want a lot of people dead.

This is not to say that older forms of terrorism have vanished. They have not. Groups motivated by nationalist/separatist agendas, left-wing ideology, right-wing ideology, and a wide range of single issues all persist and indeed even thrive. Reports by Europol, for example, highlight that 92 per cent of all terrorist attacks in the European Union between 2006 and 2008 were carried out by nationalist separatist organisations such as ETA or dissident Irish Republicans. Just six attacks or attempts were carried out by jihadi extremists such as Al Qaeda or one of its affiliates. This represented just 0.4 per cent of the total number seen in the European Union in a three-year period. A breakdown of the figures is provided in Table 1.1 below, which shows that left-wing extremists also considerably outperformed the jihadi extremists by carrying out 104 attacks.

Given such statistics, one may reasonably ask why then is the West so obsessed with jihadi extremism in particular? The answer to that takes us back to the previous point made regarding lethality. While jihadi extremists may carry out very few attacks in the West (comparatively speaking), those they do carry out are exceptionally lethal by the standards of the other terrorist groups. Between 2004 and 2010, even though jihadi extremists carried out less than one per cent of all terrorist attacks in Europe, they still managed to kill over 250 people and injure thousands more. These casualties far exceeded those caused by all of the other terrorist groups *combined*. It is the sheer lethality of these attacks, rather than their frequency, that gives analysts and security professionals pause for thought.

The ideology that drives much Al Qaeda-related terrorism sees the world only in terms of followers or 'infidels', and this legitimises an almost open-ended category of targets (Hoffman, 2006). Indeed, it is seen as a religious duty to perpetrate mass casualties. This has implications for any event that hosts large

Table 1.1 Terrorist attacks (failed, foiled or successfully executed) in the European Union

	2006	2007	2008	Total
Total	498	583	515	1,596
Nationalist/separatist	424	532	397	1,471
Left-wing	55	21	28	104
Single issue*	–	1	5	6
Islamist	1	4	1	6
Right-wing	1	1	–	2

Sources: TE-SAT 2007; TE-SAT 2008; TE-SAT 2009.

Note
*Usually animal rights and environmental groups.

concentrations of people. Thus, not only would the Olympics provide the perfect vehicle through which to transmit a message to a global audience, but it may also provide the means through which to perpetrate a mass casualty attack. The nature of the current threat has therefore provided an ominous and very different type of problem for the authorities to have to deal with.

The second reason for the importance of the timing of such a volume is that events in the international arena have clearly had an impact on the domestic terrorist threat faced by the United Kingdom in particular, and this adds to the symbolic significance of London 2012 as a terrorist target. The UK has supported much of US foreign policy, and in particular its military campaigns in Iraq and Afghanistan. Of course, the threat from Al Qaeda existed some years before 9/11 and the subsequent invasions of these two countries – evident, for example, through the African embassy bombings of 1998. But there is little doubt that events in the international arena have increased the terrorist threat to the UK. It is precisely theatres like the Middle East, Iraq and Afghanistan that are used to try to motivate recruitment to bin Laden's global jihad and to inspire attacks against countries in the West. London has, of course, already been victim to the new breed of international terrorism when, in July 2005, for the first time, suicide bombers attacked the London transport network, causing the deaths of 52 people – and the same type of threat persists. Within the UK alone, MI5 warned in November 2007, there were as many as 2,000 people involved in terrorist-related activity and MI5 suspected that there were as many again that they were not aware of (MI5, 2007), although in an interview in January 2009, the organisation's Director General spoke of significant progress against Al Qaeda that was having a 'chilling effect' on the movement (Evans, 2009). Nevertheless, in January 2010, the Joint Terrorism Analysis Centre raised the terrorism threat level from 'substantial' to 'severe', serving as a reminder of the dangers of complacency in the face of the contemporary threat (MI5, 2009).[1]

In summary, the editors believe that the contemporary relevance of such a volume is underlined by: the enormous publicity potential that elevates the Olympics as an attractive target for terrorists; the aim of international terrorists to cause mass casualties and the implications this has for mass events; London's increased profile as a target in the context of the current international environment; and the large numbers in the UK who are willing either to cause or to facilitate mass-casualty attacks within the UK homeland. Investigation into previous research on terrorism and major event security has revealed a further reason for this book – that there is, perhaps surprisingly, a paucity of literature on the subject. A number of papers addressing the broader contextual area of Olympic security do exist, although these are often specific to one particular event, such as the studies of *inter alia* Munich (Aston, 1983; Reeve, 2001), Los Angeles (Charters, 1983), Atlanta (Buntin, 2000), Sydney (Sadlier, 1996; Thompson, 1996), Salt Lake City (Decker *et al.*, 2005; Bellavita, 2007) and Beijing (Yu *et al.*, 2009). Other studies seek to undertake a more longitudinal analysis of Olympic security. Normally, such 'histories' often involve a cursory skip through Olympic 'threats' accenting Munich and Atlanta where 'events' occurred solely within the duration of the Games (*inter alia* Gamarra, 2009). Better studies (*inter alia* Sanan, 1996a, b; Thompson, 1999; Cottrell, 2003; Atkinson and Young, 2005, 2008; Hinds and Vlachou, 2007) adopt a more systematic approach, although their analyses are often much broader than terrorism-related issues. Most recently, more critical work has emerged (*inter alia* Boyle and Haggerty, 2009; Coaffee and Fussey, 2010; Giulianotti and Klausner, 2010) seeking to apply conceptual and theoretical frameworks to understanding the area of Olympic security, although this field is nascent. Other research directions have sought to understand the way terrorist threats have impacted on factors such as media reporting (Atkinson and Young, 2002) and Olympic tourism (Taylor and Toohey, 2007). Whilst valuable studies in their own right, terrorism remains peripheral to their central analyses.

The current volume

This volume is not intended to be a one-off simply because the London Olympics is on the horizon, though it will of course be of value to those practitioners and academics involved with or interested in the 2012 Games. It aims to provide readers with a useful contribution to the literature on major event security in general, to outline some of the implications of the current and projected security environment for major event security, and some of the lessons identified and learned to date. In other words, it will serve as a useful reference point not just for London 2012 but also for subsequent major events in the years ahead, both in the UK and overseas.

Drawing on both academic and practitioner input and expertise, the book itself is divided into three sections. The first is concerned with the nature of the terrorist threat. Anthony Richards, in his chapter, provides a historical context as to how sport in general has been subject to political (and military) exploitation, before assessing how and why the Olympics might be targeted by terrorists. Included in this is an assessment of the threat posed by dissident Irish Republicans who have become increasingly active at the time of writing. Finally, he considers whether or not the attack on the Sri Lankan cricket team in March 2009, the bomb attack on spectators at a volleyball match in Pakistan on 1 January 2010, and the attack on the Togo national football team just one week later represent a trend where sportspeople and/or sports events may increasingly become the object of terrorist interest as soft but high-profile targets.

Afzal Ashraf (in Chapter 3) provides a detailed assessment of the ideological orientation of Al Qaeda and how the use of terrorism is justified by exploiting religious texts. He draws a distinction between three layers of the contemporary Al Qaeda threat – Al Qaeda core, Al Qaeda associates and Al Qaeda-inspired groups and individuals. He argues that, just because no significant plot was discovered against the two post-9/11 summer Olympics of Athens and Beijing, this does not mean to say that London 2012 might not be earmarked for attack, not least because of the UK's role in both Iraq and Afghanistan and its historic relationship with Israel. Noting that attacking sports events has in fact been part of

Global Jihadi doctrine, and warning of the emergence of new Al Qaeda affiliates, Ashraf assesses the factors that may or may not prompt Al Qaeda to plan an attack on the London Olympics.

Any serious effort to combat or prevent terrorist attacks has to start by considering how terrorist groups select their targets, plan and prepare operations, and then finally attempt to carry them out. Andrew Silke analyses (in Chapter 4) the factors that impact on terrorist target selection and the decision-making and activities that go on within terrorist groups in the run up to an actual attack. The different roles that ideology, resources and experience play in this process are all explored, as is the impact of the wider security environment. In particular, this chapter examines how terrorists respond and react to security measures that have been introduced to prevent attacks, and Silke highlights some of the serious implications this has for security planners in the context of the Olympics and other major events.

Part II explores three important elements in delivering a 'safe and secure' Olympics: transport security, the role of surveillance and designing stadia for safer events. The transport system is vital for any city hosting the Olympics, thus any attack that shuts down or disrupts the transport network has a direct impact on the Games and is arguably an attack on the Olympics itself. In his chapter, Steve Swain considers various options for enhancing the security of four modes of transport: road, rail, air and maritime. He explores some of the major threats facing each of these sectors before considering options for intercepting suicide terrorists intent on attacking transport networks.

A key feature of Olympic security is the use of technological surveillance, a strategy that has become increasingly central to securing large sporting events in the post-9/11 era and one that fits neatly with the IOC's demands to prioritise the sporting event over the policing spectacle. With the London Games likely to become the first biometric and wireless Olympics, the chapter by Pete Fussey examines the way technological surveillance has been applied between the 1976 Montreal and 2008 Beijing Olympiads, and considers the implications of these processes and practices for 2012. In doing so, the range of surveillance strategies, from first-generation CCTV systems to second-generation 'video analytics', are identified before critically assessing their impact, efficacy and legacy.

Security planning to minimise the prospect and impact of terrorist attacks has also become increasingly important to the design of stadia, and now forms a key requisite for candidate cities to be awarded the right to host major sporting events. Drawing from ongoing research into the safety features built into the architecture of Olympic sites, Jon Coaffee identifies and examines key strategies adopted to augment such infrastructures. In doing so, a number of dilemmas facing organisers and security professionals are identified, alongside a consideration of the relative merits of such strategies and potential measures to augment their success.

Part III is concerned with planning, institutions, coordination, roles and responsibilities. In Chapter 8, Will Jennings provides a comprehensive analysis of the importance of threats and hazards in the organisation and staging of the

Olympics. In doing so, he draws upon wider debates about the rise of risk-management in the public and private sectors into the processes of Olympic planning. A key argument of this chapter centres on the presence of particular biases and errors commonly embedded within Olympic decision-making that impact upon the bidding for, organising and staging of the Games that can destabilise key planning and operational processes. In particular, these difficulties are made more intractable through the increased transfer of risk to markets, growth in the use of technological solutions and the spread of regulation and risk-management since the 1970s. In sum, these shifts in the ownership, management and overall governance of risk have contributed to a more complex and diverse organisational environment that has had profound effects upon the administration of the Games and its response to security and other risks.

The London 2012 bid document stated that 'private security companies ... will have an important role in public safety and security operations'. In Chapter 9, David Evans outlines and examines the role of the private security industry in providing security for the London 2012 Olympic and Paralympic Games. The key public players charged with supplying Olympic security are mapped and an anatomy of private security provision for the Olympics is provided. The chapter then examines the varying requirements for security provision in the lead-up and actual staging periods of the Games, before drawing out some key themes, such as issues of cohesion both across private-sector suppliers and, also, with public-sector agencies, as well as the need for all partners to recognise the boundaries of the private-security industry's capacity.

For any host nation attempting to 'secure' the Olympics, cooperation and coordination between different agencies (who have different operating cultures) tasked to protect major events can be one of the most formidable of challenges. The London Olympics is expected to be the largest ever policing challenge that the UK has faced. Keith Weston, in his chapter, provides a useful overview of the organisational structures created to manage the security of the Olympics of 2012, before outlining what he sees as challenging issues in multi-agency responses to crises and disasters: communication; leadership; logistics and resource management; public relations; and, finally, planning, training and exercising. Arguing that it is vital that we not only identify lessons but also learn from them, he suggests that the 'challenges of multi-agency coordination will be multiplied many times by the magnitude of the Olympics'.

Frank Gregory's chapter broadens the discussion by introducing European perspectives on major event security. He provides a background to the EU's approach to major event security in the context of the football major events security system, and discusses general information-sharing on security concerns related to major events, the specifics of EU Olympics security policy itself (including the drawing up of the *Handbook for the Co-Operation Between Member States to Avoid Terrorist Acts at the Olympic Games and Other Comparable Sporting Events*) and the issues surrounding public–private partnerships.

The collection concludes with a review of the key themes surrounding both the terrorist threats to the Olympics and the responses to these threats that have been raised in this volume. The legacy of these issues to future Olympics and other major sporting events is considered. Given that mega-event security planning is becoming increasingly standardised (across types of event and across national boundaries) at the same time that terrorist threats are continually adopting international forms, it is argued that these findings will have resonance across time and place. In this sense, issues of legacy extend beyond East London's host geography and, crucially, also apply to security planning for future mega sporting events across the globe. The volume finishes with Pete Fussey's chronology (Chapter 13) of some of the key risks and security challenges that have faced Olympic planners since the 1972 Games.

Note

1 Despite these variations, 2012 security planning has been predicated on (the current) 'severe' level of terrorist threat.

References

Aston, C. (1983) *A Contemporary Crisis: Political Hostage-Taking and the Experience of Western Europe.* Westport, CT: Greenwood Press.

Atkinson, M. and Young, K. (2002) 'Terror Games: Media Treatment of Security Issues at the 2002 Winter Olympic Games', *Olympika: The International Journal of Olympic Studies*, 9, 53–78.

Atkinson, M. and Young, K. (2005) 'Political Violence, Terrorism, and Security at the Olympic Games', in K. Young and K.B. Wamsley (eds) *Global Olympics: Historical and Sociological Studies of the Modern Games.* Amsterdam: Elsevier, 269–294.

Atkinson, M. and Young, K. (2008) *Deviance and Social Control in Sport.* Champaign, IL: Human Kinetics.

Bellavita, C. (2007) 'Changing Homeland Security: a Strategic Logic of Special Event Security.' *Homeland Security Affairs*, 3, 1–23.

Boyle, P. and Haggerty, K. (2009) 'Spectacular Security: Mega-Events and the Security Complex.' *International Political Sociology*, 3, 257–274.

Buntin, J. (2000) *Security Preparations for the 1996 Centennial Olympic Games (B) Seeking a Structural Fix.* Cambridge, MA: Kennedy School of Government, Harvard, Case Program, C16–00–1589.0, 17.

Charters, D. (1983) 'Terrorism and the 1984 Olympics', *Conflict Quarterly*, Summer, 37–47.

Coaffee, J. and Fussey, P. (2010) 'Olympic Security and the Threat of Terrorism', in J. Gold and M. Gold (eds) *Olympic Cities.* Abingdon: Routledge.

Cottrell, R.C. (2003) 'The Legacy of Munich 1972: Terrorism, Security and the Olympic Games', in M. De Moragas, C. Kennett and N. Puig (eds) *The Legacy of the Olympic Games, 1984–2000.* Lausanne: International Olympic Committee, 170–178.

Decker, S., Greene, J., Webb, V., Rojeck, J., McDevitt., Bynum, T., Varano, S. and Manning, P. (2005) 'Safety and Security at Special Events: the Case of the Salt Lake City Olympic Games', *Security Journal*, 18(4), 65–75.

Evans, J. (2009) *Director General Gives Interview to Newspapers.* Online, available at: www.mi5.gov.uk/output/news/director-general-gives-interview-to-newspapers.html.

Gamarra, A. (2009) 'Securing the Gold: Olympic Security from a Counter-Terrorist Perspective', in M.R. Haberfeld and A. von Hassell (eds) *A New Understanding of Terrorism: Case Studies, Trajectories and Lessons Learned.* New York: Springer.

Giulianotti, R. and Klauser, F. (2009) 'Security Governance and Sport Mega-Events: Toward an Interdisciplinary Research Agenda', *Journal of Sport and Social Issues*, 34(1), 49–61.

Hinds, R. and Vlachou, E. (2007) 'Fortress Olympics – Counting the Cost of Major Event Security', *Jane's Intelligence Review*, 19(5), 20–26.

Hoffman, B. (2001) 'Terrorism and Counter Terrorism after September 11th', available at: http://www.iwar.org.uk/cyberterror/resources/threat-assessment/pj63hoffman.htm.

Hoffman, B. (2006) *Inside Terrorism* (2nd Edition). Columbia: Columbia University Press.

International Olympic Committee (2010) *Olympic Marketing Fact File* (2010 edition). Online, available at: www.olympic.org/Documents/IOC_Marketing/IOC_Marketing_Fact_File_2010%20r.pdf.

MI5 (2007) *Intelligence, Counter-Terrorism and Trust.* Online, available at: www.mi5.gov.uk/output/intelligence-counter-terrorism-and-trust.html.

MI5 (2009) *Threat Level Raised.* Online, available at: www.mi5.gov.uk/output/news/uk-threat-level-raised.html.

Reeve, S. (2001) *One Day in September: The Full Story of the 1972 Munich Olympics Massacre and the Israeli Revenge Operation 'Wrath of God'.* New York: Arcade Publishing.

Sadlier, D. (1996) 'Australia and Terrorism', in A. Thompson (ed.) *Terrorism and the 2000 Olympics.* Canberra: Australian Defence Studies Centre.

Sanan, G. (1996a) 'Olympic Security Operations 1972–94', in A. Thompson (ed.) *Terrorism and the 2000 Olympics.* Sydney: Australian Defence Force Academy.

Sanan, G. (1996b) *Olympic Security 1972–1996: Threat, Response and International Co-Operation*, unpublished PhD Thesis, University of St. Andrews.

Silke, A. (2001a) 'When Sums Go Bad: Mathematical Models and Hostage Situations', *Terrorism and Political Violence*, 13(2), 49–66.

Silke, A. (2001b) 'Chasing Ghosts: Offender Profiling and Terrorism', in D. Farrington, C. Hollin and M. McMurran (eds) *Sex and Violence: the Psychology of Crime and Risk Assessment.* London: Harwood, 242–258.

Silke, A. (2003) 'Retaliating Against Terrorism', in A. Silke (ed.) *Terrorists, Victims and Society: Psychological Perspectives on Terrorism and its Consequences.* Chichester: Wiley, 215–231.

Taylor, T. and Toohey, K. (2007) 'Perceptions of Terrorism Threats at the 2004 Olympic Games: Implications for Sport Events', *Journal of Sport and Tourism*, 12(2), 99–114.

Thompson, A. (1996) *Terrorism and the 2000 Olympics.* Canberra: Australian Defense Studies Centre.

Thompson, A. (1999) 'Security', in R. Cashman and A. Hughes (eds) *Staging the Olympics: the Event and its Impact.* Sydney: New South Wales Press.

Wolff, A. (2002) 'When the Terror Began', *Time*, August 25. Online, available at: www.time.com/time/europe/magazine/2002/0902/munich/index.html.

Yu, Y., Klauser, F. and Chan, G. (2009) 'Governing Security at the 2008 Beijing Olympics', *The International Journal of the History of Sport*, 26(3), 390–405.

Part I
The terrorist threat

2 Terrorism, the Olympics and sport
Recent events and concerns for the future

Anthony Richards

Introduction

The link between terrorism and sport is not a new development (it was most dramatically evident at the Olympics of 1972 and 1996). Yet, recent events seem to indicate that sports and sportspeople have become increasingly targeted by terrorists. Examples have included the gun attack in March 2009 in Lahore on the Sri Lankan cricket team, the suicide-bomb attack on spectators at a volleyball match in Laki Marwat, north-west Pakistan in January 2010, and the gun attack in Cabinda, Angola, on the Togo national football team in the same month. Earlier, in August 2009, the English badminton team withdrew from the World Championships in Hyderabad due to a specific terrorist threat. Given the enormous publicity potential of an attack on the Olympics, the 2012 Games again raises the spectre of the convergence of terrorism and sport.

The following will assess the utility of the Olympics and sports events (and, indeed, sportspeople) as terrorist targets. Although the Olympics in general presents a 'hardened' target, given the global audience that is focused on the Games, and the fact that some venues, cities and transport systems will inevitably be less protected than others, it still represents an alluring target. Indeed, any attack on the UK during Games time would arguably be an attack on the Olympics itself. Beyond the Olympics, the following will consider whether or not the recent spate of terrorist attacks on sports represents part of a developing trend. It will ultimately argue that terrorism has had the most impact on sport (mainly cricket) in South Asia, particularly Pakistan and India, both due to direct attacks on sports targets but also due to other terrorist attacks (such as Mumbai) that have impacted on perceptions of security (or lack of) at sporting venues in these countries. The direct sporting attacks of Lahore and Laki Marwat (and indeed the attack on the Togo football team) were arguably motivated by relatively local agendas and have not since been replicated elsewhere. Nevertheless, given the propensity of Al Qaeda to target high profile and mass-casualty targets, and the indiscriminate nature of its violence permitted by its ideological parameters, it would be dangerous to ignore the potential threat that it represents to major sports events and sportspeople in Western countries and elsewhere.

This chapter will begin by briefly assessing what is meant by 'terrorism' and how some features of terrorism illuminate the significance of events such as the Olympics for terrorist targeting purposes. Terrorism is communication and the Olympics provides a prime opportunity and illustration of the value of the media in publicising a terrorist organisation's cause. Before assessing the utility of such events as targets for terrorists, however, the chapter will provide a brief historical overview as to how in fact the Olympics and sport in general have more broadly been subject to political manipulation, dating back to ancient times when they were exploited to enhance physical strength and military prowess. It will then go on to consider how they have been politically exploited by those using terrorism and what impact this has had on sport, before returning to the Olympic theme in particular by considering the threats that Al Qaeda and dissident Irish Republicans pose to London 2012.

Features of terrorism and their implications for the Olympics

Terrorism can be described as a particular method of violence and/or the threat of violence that has been carried out by a wide range of actors (both state and non-state), that often targets civilians, is usually for a political purpose and is usually designed to have a psychological impact beyond the immediate victims. A subset of terrorism is the concept of non-state terrorism, which can be described as a particular method of violence and/or the threat of violence that has been carried out by non-state actors in order to coerce a government and its population, that often targets civilians, is usually for a political purpose and is usually designed to have a psychological impact beyond the immediate victims. While states are most certainly culpable as perpetrators of terrorism (whether they have carried out acts of terrorism, have sponsored terrorism or have perpetrated what has often been called 'state terror'), it is non-state terrorism that states are primarily concerned with in their efforts to defend their homelands.[1] This distinction is important because it allows us to transcend the current debate over the extent, if at all, to which 'the state' as perpetrator has been excluded from terrorism studies (and also from definitions of terrorism) (Jackson, 2009). So for the purposes of this chapter, the definition of non-state terrorism will be used.

Notwithstanding such distinctions, terrorism seeks to generate a wider impact beyond the immediate victims of an attack. It is about communicating a message, often to a number of target audiences (whether they be a perceived constituency of support, the international community or an adversary) – and it is the desire to transmit a message at the global level and even to capture the world's attention (as the Black September attack did in 1972) that potentially makes sports events an attractive target.

The Olympic movement, as an international endeavour that seeks to promote peace and harmony, is ideally positioned to appreciate one of the major obstacles to countering terrorism – the inability of the international community to agree a definition of the phenomenon and therefore the degree of difficulty in generating international cooperation against those who would resort to terrorism. Even after the most devastating attack ever to have afflicted the Olympics, in Munich 1972, the then Secretary General of the United Nations, Kurt Waldheim, in his efforts to encourage member states to agree to the need for 'measures to prevent terrorist and other forms of violence which endanger or take human lives or jeopardise fundamental freedoms', was confronted by deep concern from some African and Arab states that those engaged in legitimate national liberation struggles would be classified as terrorists (Wardlaw, 1989: 105). Although it seems to be more useful to view terrorism as a *method*[2] as opposed to focusing on those definitions that are perpetrator-based, the problem of the subjective nature and use of the term persists to this day.

Sport as a political and military weapon

Before exploring further the exploitation of the Olympic Games by terrorist groups, it is worth reminding ourselves that the Olympics and sport in general have long been the subject of political exploitation. Despite calls for politics to be left out of sport (perhaps most notably by those who advocated the increase of sporting links with South Africa during Apartheid), the fact is that sport has inevitably become prey to political manipulation both in the contemporary world and indeed in ancient times.

'Sport' is a shortened form of the original term of 'disport', meaning a diversion or an amusement (Brasch, 1986: 1). Apart from the natural desire to compete, the impetus behind sport was also for man to be able to more effectively defend himself and his tribe, to learn to run fast, to jump and to swim. Sporting activities such as archery, judo and karate were invented in order to tackle opponents (ibid.: 2). The development of sports that required physical strength found much of its origin in the desire to defend against and conquer one's enemies. Thus chariot racing, boxing and wrestling were inextricably linked to the defence of cities; indeed, 'ultimately, there was only one intent and aim of athletic contests: to feint the stress of battle; to stay sharp and ready for war' (Spivey, 2004: 18). Despite the not uncommon contention that 'good athletes do not necessarily make good soldiers', physical strength and traditions of 'rigorous gymnastic upbringing', perhaps typified by the Spartans, were linked to military prowess (ibid.: 27). Miller, referring to the Greek city-states from around 525 BC, states that 'athletics were clearly a tool of ancient political aggrandizement – just as they are today', with coins used to 'advertise victories at Olympia and other games' (Miller, 2004: 218).

The intertwinement of military power and Olympic success was illustrated by the entry by Alkibiades of seven chariots at the Olympiad of 416 BC against the backdrop of the Peloponnesian War. In anticipation of an Athenian victory, he apparently proclaimed:

> The Greeks who had been hoping that our city was exhausted by the war came to think of our power as even greater than it is because of my magnificent embassy at Olympia. I entered seven tethrippa [chariots], a number

never before entered by a private citizen, and I came in first, second, and fourth.

(Miller, 2004: 221)

Many centuries later, the link between sports and the military was also explicitly encouraged by the United States armed forces in the run up to the First World War, where sports and athletic training were made 'a central component of military life' (Pope, 1997: 139). Indeed, sports were accepted as 'essential elements of a soldier's responsibility' that 'made good military sense in developing needed physical endurance' (ibid.: 144).

Sport has also been used to promote national fervour and nation-building. Italy's football World Cup triumph in 1938 was seen as uniting the Italian diaspora behind Mussolini's fascist regime and for generating a communal identity (Martin, 2004: 1). To this end the regime was credited with regenerating Italian society through sport, with particular emphasis on exploiting the mass appeal of football (or 'calcio') for political goals (ibid.: 15, 2). The 1936 Olympics in Berlin, which was awarded to the city at least partly to bring Germany back into the international fold after the First World War, provided Adolf Hitler with an opportunity to showcase the Germany that he was moulding, and it became an enormous propaganda exercise for nazism (Hilton, 2006). Apart from such nation-building, the Olympic Games has also been the victim of the turmoil of international politics. They did not take place at all in 1916, 1940 and 1944 because of the two world wars.

The very fact that athletes are identified with a state is a political statement in itself. Indeed, it was hoped that through the presence of their delegation at the Atlanta Games, Palestinians could use sport as a stepping stone to nationhood (Sanan, 1997: 305). For host nations it is seen as an opportunity to showcase their country to the many thousands of tourists and millions of spectators. It can therefore also, conversely, be used as a political tool to undermine a host country's attempts to enhance its international profile and, as such, the spectacle has often been subject to the vagaries of the international politics of the time. The context of the Cold War, for example, provided the backdrop against which boycotting the Olympics became a political weapon. The US-led boycott of the Moscow Games in 1980 (in response to the Soviet invasion of Afghanistan) was reciprocated by the Soviet Union when it stayed away from the Los Angeles Games of 1984.

Sports have been exploited for other, more positive, political reasons – as the means to underpin diplomacy in international politics. This was perhaps most evident through the 'ping-pong' diplomacy that took place between the United States and China after the American 'Ping-Pong' team accepted an invitation to visit China in April 1971. It was followed by the visit of President Nixon to the People's Republic in 1972. In this case, sport had clearly served as a lever to improve relations between the two countries.

Yet, while sports events have been manipulated for a variety of reasons, conversely, 'all sports have [also] had to take a view on international politics' (Hill,

1996: 34). Hill uses the example of the bridge world championships, where Israel and its 'numerous' enemies were drawn in separate qualifying pools in the hope that they would not progress far enough to ultimately have to play against each other (ibid.). The Gleneagles Declaration of Commonwealth nations in opposition to apartheid discouraged athletes and sportsmen from competing against their South African counterparts. This confirmed their earlier commitment in 1971 to oppose racism and was followed by their Declaration on Racism and Racial Prejudice in 1979. This is in line with one of the IOC's fundamental principles of 'Olympism', which is: 'any form of discrimination with regard to a country or a person on grounds of race, religion, politics, gender or otherwise is incompatible with belonging to the Olympic Movement.' Hence the suspension of South Africa from the Games in 1970, which explains its absence from 1972–1988 inclusively.

The Ancient Games and the *ekecheiria*

Much is made of the peaceful mission of the Olympic Games and how terrorism flies in the face of the very ideals that the Olympics stands for. In ancient times, the many different autonomous Greek regions laid down their arms to 'enjoy the "divine peace" associated with the Olympic competitions' in honour of Zeus (Sinn, 2000: 1). Hostilities ceased for a one-month period, such was the high regard held for the Games which were treated as inviolable. What was known as the *ekecheiria* was 'an indispensable precondition for continuous and undisturbed holding of the Olympic Games', and it was seen not as an arbitrary human institution but as divine law (Lammer, 2010: 8). This was because 'the area of the sanctuary of Zeus at Olympia was declared holy (Greek: hieros) and therefore sacrosanct', and anyone breaking these rules would have to fear the revenge of Zeus (ibid.: 4). As such, athletes, their coaches and spectators could travel freely to and from Olympia without fear of being attacked or robbed, even if they were travelling through land governed by those against whom their own rulers were at war. Thus, quite in contrast to the Olympics being seen as an opportunity for terrorists to exploit through violence or being cancelled because of war (1940, 1944), the Games in ancient times were highly respected and a truce and temporary respite from wars took place in their honour, though there were exceptions (see below, and Lammer, 2010: 11–12). This did not mean, however, that the Olympics was immune from becoming 'a field of propaganda for the great powers and ... a continuation of war by other means' (ibid.: 16).

Not that the Games of the period were entirely exempt from political violence. In 364 BC the Eleans attacked the Olympics hosted by the Arkadians, before severe losses forced the former to retreat. Henceforth they labelled the games as an Anolympiad (Non-Olympiad) (ibid.: 222). Later, in 235 BC, against a backdrop of political conflict between Argos and Aratos of Sikyon, and after the Nemean Games had been moved to Argos, a rival Games was held at Nemea. Aratos, whose decision it was to launch the rival Games at Nemea, 'captured and sold into slavery' any athlete he caught travelling through his territory to the Games at Argos (ibid.: 222).

Nevertheless, Miller points out that whatever political manoeuvrings took place, including the rare cases of political violence, during the more than a millennium of ancient Greek Olympic Games, the 'games went on'. This, he argues, is in contrast to the past century of modern Olympics where Munich 1972 and three major boycotts (1976–84) seriously threatened the events (although these Games did 'go on') and where the Games of 1916, 1940 and 1944 did not even take place at all because of the political environment (Miller, 2004: 225).

Notwithstanding these subsequent interruptions, one of the core motivations for Baron de Coubertin in reviving the modern Olympic Games towards the end of the nineteenth century was 'for international understanding and peace' (Barney, 2007: 225). The Olympic Charter describes 'Olympism' as seeking to 'create a way of life based on the joy of effort, the educational value of good example and respect for universal fundamental ethical principles' (IOC Charter: 11). It states that: 'The goal of Olympism is to place sport at the service of the harmonious development of man, with a view to promoting a peaceful society concerned with the preservation of human dignity' (IOC Charter: 11). And some have gone so far as to suggest that 'Olympism is a philosophy of life which uses sport as a conveyor belt for its ideas' (Hill, 1996: 258). Clearly, any attack, therefore, on the Games flouts these ideals. Yet, this does not necessarily mean that the terrorists who have targeted or have aimed to target the Games have done so deliberately and specifically to oppose these values. More accurately, these Olympic aspirations are of relative insignificance compared to the objectives of terrorist organisations and the enormous potential that the Olympics provides in relation to advertising their cause to a vast international audience.

Terrorism and the targeting of sports

To use the non-state definition of terrorism referred to above: it is a particular method of violence and/or the threat of violence carried out by non-state actors in order to coerce a government and its population, that often targets civilians, is usually for a political purpose and is usually designed to have a psychological impact beyond the immediate victims. To reiterate, terrorism is therefore a form of communication that is designed to send a message to a wider audience beyond the immediate victims. The media, including the Internet, provide the means through which the message can be delivered, while the prestige and popularity of the Olympics provides the stage with its audience of millions. The Olympics is one of the most major global media events of all time and any attack by terrorists is designed to, and almost certainly would, attract unprecedented publicity for the cause of the perpetrators.

George Habash, the leader of the Popular Front for the Liberation of Palestine, was quoted as writing that:

a bomb in the White House, a mine in the Vatican, the death of Mao Tse-Tung, an earthquake in Paris could not have echoed through the consciousness of every man in the world like the operation at Munich.... The choice of the Olympics, from the purely propagandist viewpoint, was 100 per cent successful. It was like painting the name of Palestine on a mountain that can be seen from the four corners of the earth.

(cited in Taylor, 1993: 6)

One of the organisers of the Munich terrorist attack argued that 'we have to kill their most important and famous people. Since we cannot come close to their statesmen, we have to kill artists and sportsmen' (Hoffman, 1998: 71). This statement implies that the theories of displacement targeting discussed elsewhere in this volume (Silke, see Chapter 4) apply as far back as 1972 – that as more obvious targets associated with the enemy state become more protected, then 'softer' targets may be selected (at a time when the Olympics did not represent such a 'hardened' target). More recently, for example, A US RAND report noted that in mid-2004 the South East Asian group Jemaah Islamaya had apparently concluded that targeting 'well-protected targets was beyond its capabilities and instead opted for attacks on relatively unprotected soft targets' (RAND, Vol. 2: 73). So, in relation to sports events and people, one has to consider in the post-9/11 world (where more obvious targets may be well protected) that alternative targets may be chosen. With the exception of the Olympics, where its main venues are likely to be well protected, sports and sportspeople more broadly may become targets as alternative and 'softer' options.

Indeed, sports beyond the Olympics have been the object of terrorist attention before. It was widely believed that the IRA was behind the kidnapping of the Derby winning racehorse Shergar in 1983, and the same group was responsible for coded bomb warnings that resulted in the abandonment of the Aintree Grand National in 1997. In May 2002, ETA detonated a car bomb outside the Bernabau stadium just before a Champions League football match between Real Madrid and Barcelona, and in 2008 a suicide bomber attacked the start of a marathon outside Colombo, killing 12 people including a former Olympic marathon runner and a national athletics coach. In January 2008, it was reported that the Dakar rally was cancelled because of 'direct' threats of terrorism (*Guardian*, January 2008), while the same reason was cited for the withdrawal of the English badminton team from the World Championships in Hyderabad in August 2009. In March 2009, six policemen and a bus driver were killed and seven cricketers and a coach were injured when the Sri Lankan cricket team's convoy was attacked by gunmen in Lahore, while on 1 January 2010 nearly 100 people were killed when a suicide bomber attacked a volleyball match in Laki Marwat, Pakistan. A week later, the Togo national football team was attacked by gunmen, killing at least two and injuring seven. In April 2010, explosive devices were planted outside Chinnaswamy cricket stadium in Bangalore shortly before an Indian Premier League match.

A distinction should be drawn between those terrorist attacks that directly target sports events and people (such as Lahore) and those that don't but that do have broader security implications for the safety of such events (such as Mumbai). Terrorism in general has had a serious impact on cricket in Pakistan and its prospects of hosting touring teams. The Mumbai attacks that lasted from 26–29 November 2008, killing more than 170 people, and that captured the attention of the world's media, was followed by the withdrawal of India from a planned cricket tour there in January 2009. Pakistan also lost out on hosting the 2009 International Cricket Council Champions Trophy. For many, the Lahore attacks only served to confirm Pakistan as an insecure place to play cricket. South Africa also withdrew from a scheduled October/November 2010 tour there and, at the time of writing, there seems little prospect of any international cricket taking place in Pakistan in the foreseeable future. This has left its cricket team with the only option of playing its 'home' games in neutral venues.

The impact of terrorism on sport has been felt in India as the Mumbai attacks also called into question the safety of India as a cricketing venue, with the prestigious Indian Premier League having to be relocated to South Africa in April 2009, while after the Lahore attack on a sporting target, there have been concerns for the safety of the Commonwealth Games in India in October 2010. After Mumbai, a leading English cricket commentator proclaimed, due to the subsequent 'oppressive security measures taken in the hotels and cricket grounds in Chennai and Mohali before Christmas [2008]', that 'if we're honest we know terrorism's won in India' (Agnew, 2009). Aside from the direct attack on its cricketers at Lahore, Sri Lankan cricket had also suffered some years earlier (in 1996) when a bomb attack in the capital of Colombo prompted Australia to withdraw from a World Cup cricket match. Further afield, and beyond cricket, golf also became the victim of terrorism when the United States Ryder Cup team postponed its golf match with Europe in the aftermath of the 9/11 attacks.

The targeting of sports – a developing trend?

Although there has been a spate of terrorist attacks on sports recently, it is difficult to argue that it represents part of a developing trend where sports events and sportspeople will increasingly be targeted by terrorists. One attack that would feature strongly (because of its high profile) in any such hypothesis of a developing trend was that carried out on the Togo national football team by the Front for the Liberation of the Enclave of Cabinda (FLEC), or a faction of it. There is, however, some doubt as to whether or not the team itself was earmarked for attack. Although such reports should be treated with serious caution, the 'Secretary General' of the group was quoted as saying that 'this attack was not aimed at the Togolese players but at the Angolan forces at the head of the convoy.... So it was pure chance that the gunfire hit the players' (Sturcke *et al.*, 2010). It is, of course, difficult to verify how genuine these claims are (especially when factions of the FLEC movement have targeted foreign nationals before for kidnap), and the attack (particularly the shooting of the team's bus) undoubtedly gave enormous exposure to the group and its cause of an independent Cabinda. This does not mean, however, that it generated any sympathy for its goal – in fact the 'condolences' that were expressed to the victims' families and the Togolese government by the perpetrators might suggest that they felt the attack to be a mistake that was counterproductive to their cause. Certainly, any such attacks in the future against foreign sports teams would render such sentiments as hollow. In conclusion, it is by no means clear, and probably unlikely, that this type of attack is something that would be repeated as part of a tactical shift by the group or its offshoots.

The other recent attacks were also arguably borne of local and regional agendas. The suicide attack on volleyball spectators was said to be a response to Laki Marwat residents who had established a militia to expel militants from the area (BBC, 2010), while the Lahore attack on the Sri Lankan cricket team was the latest riposte in Pakistan's struggle to combat its extremists within. In recent times, therefore (and the Togo case notwithstanding), it appears that only South Asia has suffered from clearly deliberate attacks against sports events and people.

It is therefore premature to suggest any trend towards sports targeting on the part of terrorists, although terrorism in general has clearly had a major impact on sport in Pakistan and India in particular. One should not assume, however, that sports events, teams and individuals would not be targeted in the future and in other parts of the world, including Europe and the USA, especially given the nature of the contemporary international terrorist threat. The utility of targeting such events and people varies. Some with nationalist agendas, who may to some extent be answerable to domestic political constituencies, might refrain from attacking iconic sportspeople who have a wide following, or from causing a high number of casualties at sports events, for fear of public revulsion that would be detrimental to their cause.[3] For others, who are driven by a religious and international ideology that justifies attacking all 'infidels', such restrictions may not apply. This includes Al Qaeda, which has a broad category of targets and no aversion to causing mass casualties. Moreover, the global attention that the Lahore cricket and Togo football episodes attracted might deem such targets as appealing options for those whose ideological parameters permit such attacks. The utility of sports events as targets, then, lies in the presence of large concentrations of 'infidels', at the same time as providing the potential for widespread publicity – the higher the target's profile, then the greater the publicity. The list of potential sporting targets is endless and their security provision is variable. Quite apart from national sporting events, in the UK alone there are nearly 50 Premiership, Championship and League football matches every week during the football season, not to mention major club rugby fixtures in the Heineken Cup, the Guinness Premiership and the Magners League, or indeed county cricket championship fixtures in the summer.

Al Qaeda and the London Olympics

Beyond the fact that events like the Olympics serve to provide a global stage from which terrorists can disseminate their message, and the possibility that sports-related targets in general may become increasingly preferred as 'softer' options, one should also consider the nature of the contemporary terrorist threat and what impact this may have on the vulnerability of sports events and the Olympics in particular. Al Qaeda sees the United Kingdom as culpable for Muslim suffering, most particularly through Britain's resolute support for the United States in Iraq and Afghanistan. As such bin Laden and his deputy, al Zawahiri, have repeatedly warned the UK that it will be held to account for its role abroad. As noted elsewhere in this volume (p. 6), there are also significant numbers within Britain who would happily oblige by perpetrating attacks within the homeland.

While any terrorist attack on the Olympics is also sure to generate enormous publicity for the perpetrators, such a target might fulfil another Al Qaeda objective – the aim of causing mass casualties. Brian Jenkins famously (and, in general, correctly) declared that 'terrorists want a lot of people watching, not a lot of people dead', but it is an assertion that has increasingly been brought into question (for example, after Aum Shinrikyo's attempt to kill large numbers of people on the Tokyo subway in 1995) – and it no longer holds true for the contemporary terrorist threat inspired by Al Qaeda. Indeed, the convergence of a radical religious ideology that justifies perpetrating mass casualties and a capability of using unconventional weapons amounts to a nightmare scenario for governments.

Therefore, any heavy population centres or any event where there is a mass congregation of people or 'infidels' potentially represents a target for Al Qaeda. Thus, not only do the Olympics provide an enormous opportunity to publicise its cause but, such is the open-ended category of targets legitimised by its doctrine, it also provides heavy concentrations of people for the network to target. Indeed, Afzal Ashraf notes in this volume that the *Encyclopaedia of Afghan Jihad*, for example, has included football stadiums as suitable targets.

Beyond the publicity potential and the fact that the Games represents a mass-casualty target, there are other potential reasons for an attack on the Olympics. Much has been made recently of the view that Al Qaeda is on the back foot. In a January 2009 interview, the Director General of MI5 spoke of the progress being made in the UK against Islamic extremists that was having a 'chilling effect' and was forcing them 'to keep their heads down' (MI5 website). In his Mansion House foreign policy speech in November 2009, the British Prime Minister announced that 'methodically, and patiently, we are disrupting and disabling the existing leadership of Al Qaeda' and 'since January 2008 seven of the top dozen figures in Al Qaeda have been killed, depleting its reserve of experienced leaders and sapping its morale' (Gordon Brown, Mansion House speech, 16 November 2009). In this context, Al Qaeda may want to show that it still has the capability of launching major attacks, not least to motivate and mobilise its supporters and to counter such claims that its morale was being sapped. Much of its effort is focused in Pakistan, but an attack in the UK during Games time would send a powerful and symbolic message that it still very much has a global capability. It would also cite the cause of any attack as the British and NATO presence in Afghanistan (if the British are still there) or indeed UK involvement in any other theatre abroad, with the aim of undermining the British public's commitment to such endeavours. Thus, not only would the resolve of the British in tolerating or supporting foreign exploits be tested by deaths of British soldiers abroad, but also potentially by civilian casualties at home.

The question as to whether or not Al Qaeda will try to attack the Olympic Games of 2012 depends, of course, as to what is meant by 'Al Qaeda'. The label has been used to describe the core Al Qaeda leadership headed by Osama bin Laden and Ayman al Zawahiri, those groups affiliated to it, and those perhaps relatively autonomous individuals who want to carry out acts in the name of Al Qaeda's ideology. The level at which any attack is planned may have a direct bearing on the sophistication and type of tactics that might be used.

As a decentralised global network, it is not often clear which attacks have been planned and authorised by Al Qaeda itself and which ones have been carried out by relatively autonomous groups of individuals who want to perpetrate terrorist attacks in its name. The targeting calculus may be different for the two. This links in with an important debate that has a direct implication for the way governments might appropriately respond to the contemporary international terrorist threat and, indeed, on the level of sophistication that the authorities in 2012 may be confronted with in any terrorist attack. The discussion revolves around the belief, on the one hand, that the contemporary threat is largely composed of autonomous individuals and groups inspired by Al Qaeda and wanting to act in its name (articulated by Marc Sageman in *Leaderless Jihad*, 2008) and, on the other, that the terrorist plots that are being monitored by the security services and those that have been uncovered are more centrally directed from elsewhere (usually Pakistan) than has hitherto been acknowledged (a view put forward by Bruce Hoffman, 2008). Since this debate materialised in 2008, it appears that the latter analysis has been borne out by both security service assessments and court findings in the UK. It transpired, for example, that the transatlantic 'airline plot' was largely directed from Pakistan and the Prime Minister (citing the Director General of MI5) announced that three-quarters of the most serious plots that the security service was currently tracking have links to Pakistan (Brown, 4 September 2009).

This debate on the structure of Al Qaeda has implications both for the type of tactics that might be employed and how one might respond to the threat more broadly. If self-motivated individuals are acting autonomously in response to how they see their environments (both domestically and internationally), then policies can be adopted in the form of social programmes to address such perceptions. On the other hand, if key players are involved in recruitment and training, then an approach that seeks to apprehend these pivotal people might take priority. A combination of both approaches would seem to be sensible because, even if there is some form of direction from countries like Pakistan, clearly they are recruiting willing people.

As far as the implications for terrorist tactics are concerned, relatively autonomous groups are arguably likely to be more amateur and less organised in their approach than those who have received a degree of training, organisation and competence from experienced operators abroad. The disparate nature of the threat, then, can account for the range of attack scenarios that may emanate from what we call Al Qaeda: from the botched London and Glasgow attacks of June 2007 to the more sophisticated transatlantic bomb plot of 2006.[4] As far as the Olympics are concerned, the threat scenario includes multiple and simultaneous suicide attacks (a hallmark of Al Qaeda operations) as well as amateur attempts to act in Al Qaeda's name. Yet, even with the latter, the Olympic context would propel them into highly dramatic events. Likewise, attacks away from the Olympic site, or against the transport network, but clearly designed to disrupt the Olympics, would also receive worldwide attention. Indeed, any attack in the UK during Games time would arguably be seen as an attack on the Olympics itself. The choice of targets, however, would be highly dependent on the perceived level of security around them, with preferred targets, such as Olympic venues and certainly the Olympic Park itself, becoming less attractive as a result.

Dissident Irish Republicans

The utility of attacking the Olympic Games is, of course, likely to be different for different terrorist organisations. Al Qaeda and those adhering to its ideology have shown a propensity to cause mass casualties, hence crowded places may have a particular allure. The tactics of other, more 'traditional', terrorist groups may fall into the category of 'terrorists want a lot of people watching, not a lot of people dead'. This might apply to nationalist/separatist organisations that claim to represent a domestic constituency.

Of particular concern in this regard for the UK has been the continuing and increasing activity of dissident Republicans who have now inherited the mantle of traditional Republican ideology from the IRA. In the longer term, if the peace process does not ultimately deliver a united Ireland, it is a doctrine that is potent enough to lure more Irish Republicans to the 'physical force' tradition and its imperative of driving the British from the province through the use of violence. On 7 March 2009, Sappers Mark Quinsey and Patrick Azimkar were shot dead by the Real IRA and, according to the Independent Monitoring Commission, this 'represented a major escalation of RIRA activity' (IMC, Twenty-Second report). Two days later, the Continuity IRA claimed responsibility for the killing of PC Stephen Carroll. In September 2009, a 460-pound bomb was discovered in Forkhill, Armagh (IMC, Twenty-Second report). The Commission warned that 'both RIRA and CIRA had remained extremely active and dangerous', that they had shown 'a capability to plan and organise' and that their activity since the early summer of 2008 'had been consistently more serious than at any time since [they] had started to report in April 2004' (IMC, Twenty-Second report). In the six months between March and August 2009 'the seriousness, range and tempo of their activities all changed for the worse' (IMC, Twenty-Second report).

Dissident Republicans were responsible for the Omagh bomb in 1998 that killed 29 people and unborn twins. This was in many ways seen as an 'own goal' as the public revulsion against the attack and the perpetrators forced the Real IRA into a temporary ceasefire. Nevertheless, it did not prevent the group from targeting civilian areas once it had re-launched its campaign. One feature of this particular threat that gives cause for concern for 2012 is that Irish Republicans have traditionally regarded successful attacks on the mainland as of much greater value than those carried out in Northern Ireland, as they are believed to have a much more compelling impact on the British public. Part of IRA strategy when it was not on ceasefire was to persuade the British public through coercion to, in turn, compel the government to withdraw from the province.

The Real IRA (which split from the IRA in 1997) has also targeted the mainland, most particularly through a spate of attacks in 2000–2001, including bomb attacks on Hammersmith Bridge and Ealing. Bombs also exploded outside the BBC Television Centre in Shepherd's Bush, and outside a postal sorting office in Hendon, while a bomb failed to detonate properly in Birmingham city centre. The group also launched an audacious missile attack against the MI6 headquarters building. There were no deaths as a result of these attacks (although there were injuries) but, at least in the cases of a bomb exploding outside busy pubs at closing time in Ealing Broadway (August 2001) and the Birmingham device, this was because of good fortune rather than any intent to avoid fatalities on the part of the perpetrators. In relation to current dissident Republican activity, the IMC noted that 'in April RIRA told a newspaper that it would mount an attack in Great Britain when it became opportune to do so' (IMC, Twenty-Second report). There is no question that if the group had the capability of attacking the UK mainland then it would follow through with this ominous warning.

In theory, organisations like the Real IRA, as a nationalist/separatist organisation, would not be interested in causing mass casualties at the Olympics of 2012 but, rather, would aim to fully exploit the British Olympics for three main reasons: to generate international publicity for its cause, to prove that it is capable of carrying out such an attack, and to disrupt the British attempt to deliver a successful and safe Games. More strategically, it would want to use violence to rejuvenate the traditional Republican belief that partition in Ireland and any 'partitionist' political dispensation in Northern Ireland are inherently unsustainable. Dissident Republicans would, one can assume, derive enormous satisfaction from disrupting the Games that might otherwise have been a great source of national pride for the British government and its people. Any aim to cause as many casualties as possible, however, would prove enormously counter-productive (as Omagh showed), diminishing any prospect of increasing the little support base that dissident Republicans have, both at home and abroad in the United States.

Aside from the above considerations, one of the calculations that terrorist organisations have to make is whether or not it would be counter-productive to their cause to attack the Olympics itself (directly or indirectly). For example, one commentator, speaking of Munich 1972, suggested that it was a 'net loss' for the Palestinian cause, arguing that:

> Munich ... gave the Palestinians an image of mindless, bloodthirsty thugs, more so as the venue for the operation is considered historically a sacrosanct occasion of hope and peace. Accordingly, this put the Palestinians outside the circle of civilised humanity in the eyes of many who might otherwise have been sympathetic.
>
> (Sanan, 1997: 107)

Indeed, given the international and peaceful ethos of 'Olympism', and given that one of the aims of many terrorist organisations is to attract international sympathy for its cause (and not widespread condemnation), perhaps targeting the Olympics is not conducive to the goals of some terrorist organisations. Nevertheless, as noted above, this particular concern might be outweighed by the enormous publicity that any attack on the Olympics would generate. In this context one could perhaps conclude that, in the case of dissident Irish Republicans, they would aim to cause maximum *disruption* to the British showpiece rather than the maximum casualties that Al Qaeda has sought to perpetrate.

Terrorist tactics

Appreciating the different ideologies of terrorist organisations that might be inclined to attack the Olympic Games in turn sheds light on the type of tactics that may be used. The doctrine of Al Qaeda justifies attacks against all who are 'infidels'. Bin Laden himself explicitly stated in his 1998 fatwa that:

> the ruling to kill the Americans and their allies – civilians and military – is an individual duty for every Muslim who can do it in any country in which it is possible to do it.... This is in accordance with the words of Almighty Allah.

It is this threat and the desire to cause mass casualties, therefore, that provides the greatest danger to the London Olympics of 2012. One feature of the Olympics that lends itself to the modus operandi of Al Qaeda is that it has multiple venues. Simultaneous attacks have been one of the hallmarks of the contemporary threat and, if Al Qaeda was considering a serious and sophisticated attack on the Olympics, then this is a scenario that should be factored into security planning. A worrying development that also has to be noted is the organised gun attacks in Mumbai and on the Sri Lankan cricket team. Again, the thought of professionally trained gunmen shooting at crowds in congested areas with advanced weaponry is a concerning one, though it is hoped that security checks will be effective enough to prevent such a scenario.

If dissident Irish Republicans were to try to attack the London Olympics, it is unlikely that they would aim to cause mass casualties. Although the possibility of a bomb attack should not be discounted, their level of capability may be limited to hoax calls with the aim of achieving maximum disruption. There could, of course, also be acts of sabotage. This would not be a new tactic as far as the Olympics are concerned. At the 1992 Winter Olympic Games of 1992 in Albertville, television transmission cables were severed by a member of a radical environmentalist in order to disrupt the broadcast of the opening ceremony (Sanan, 1997: 128). There were also sabotage attacks that attempted to disrupt the opening ceremony of the 1992 Barcelona Games, one of which was carried out by the small Spanish Marxist group GRAPO (ibid.).

Conclusion

Major sports events, and especially the Olympics, provide terrorist organisations with opportunities for generating enormous international publicity for their demands. Terrorism is communication, and these events provide an avenue through which to communicate its message to as wide an audience as possible. The Olympics in general is likely to be well protected, but it still represents an attractive target for terrorists who would aim to benefit from the global exposure of any attack and to disrupt the host country's attempt to deliver a 'safe and secure' Olympics. Other sports events and sportspeople are less well protected than those during Games time and there have been a number of recent attacks and threats against such targets. Terrorism has lately had a major impact on sport in Pakistan and India, for example, but it is too premature to suggest that recent events are part of a growing trend of terrorist attacks against sporting targets in general. Nevertheless, in the face of the contemporary Al Qaeda threat and its modus operandi, there is no reason to hope that such targets will be exempt from the targeting calculus of terrorists in the months and years ahead.

Notes

1 The exception to this in the Olympic context was the threat posed by North Korea against the Seoul Games of 1988.
2 As Leonard Weinberg rightly observed, the notion of 'one man's terrorist is another man's freedom fighter' confuses the goal with the activity (Weinberg, 2005: 2).
3 They may, however, still perpetrate lower-level attacks that can still cause death and injury and may do so with the aim of achieving maximum disruption rather than mass casualties.
4 This is not to say that cells with such direction from elsewhere may not also botch their attacks, such as the one that attempted to blow up underground trains on 21 July 2005.

References

Barney, Robert K., 'The Olympic Games in Modern Times', in Schaus, Gerald P. and Wenn, Stephen R. (eds), *Onward To The Olympics, Historical Perspectives On the Olympic Games*, Wilfrid Laurier University Press, 2007.
Brasch, R., *How Did Sports Begin?*, Tynron Press, 1986.
Hill, Christopher R., *Olympic Politics, Athens to Atlanta 1896–1996*, Manchester University Press, 1996.
Hilton, Christopher, *The 1936 Berlin Olympic Games*, Stroud, 2006.
Hoffman, Bruce, *Inside Terrorism*, Cassell, 1998.

Jackson, R. et al. (eds), *Critical Terrorism Studies: a New Research Agenda*, Routledge, 2009.
Lammer, M., 'The So-Called Olympic Peace in Ancient Greece', in Konig, J. (ed.), *Greek Athletics*, forthcoming, Edinburgh University Press, 2010.
Martin, Simon, *Football and Fascism, The National Game Under Mussolini*, Berg Publishers, 2004.
Miller, Stephen G., *Ancient Greek Athletics*, Yale University Press, 2004.
Pope, S.W., *Patriotic Games, Sporting Traditions in the American Imagination, 1876–1926*, Oxford University Press, 1997.
Sageman, M., *Leaderless Jihad*, University of Pennsylvania Press, 2008.
Sanan, 'Olympic Security 1972–1996: Threat, Response, and International Cooperation', PhD Thesis, Department of International Relations, University of St Andrews, April 1997.
Sinn, Ulrich, *Olympia: Cult, Sport, and Ancient Festival*, Markus Weiner Publishers, 2000.
Spivey, Nigel, *The Ancient Olympics*, Oxford University Press, 2004.
Taylor, Peter, *States of Terror*, BBC Books, 1993.
Wardlaw, G., *Political Terrorism: Theory, Tactics and Counter-Measures*, Cambridge University Press, 1989.
Weinberg, Leonard, *Global Terrorism: a Beginner's Guide*, Oneworld, 2005.

Journals

Critical Studies on Terrorism, Routledge.
European Political Science, Volume 6 (3), Cambridge University Press.

Web references

Agnew, Jonathan, 'If We're Honest, Terrorism's Won in India', *Evening Standard* online, 5 March 2009, available at: www.thisislondon.co.uk/standard-sport/article-23657930-if-were-honest-we-know-terrorisms-won-in-india.do.
BBC News, 'Pakistan Volleyball Bomb Toll Climbs to More Than 90', 2 January 2010, available at: http://news.bbc.co.uk/1/hi/world/south_asia/8437473.stm.
Gordon Brown's Mansion House speech, 16 November 2009, available at: www.number10.gov.uk/Page21339.
Gordon Brown speech on Afghanistan, 4 September 2009, available at: www.number10.gov.uk/Page20515.
Hamilos, Paul, 'Dakar Rally Cancelled at Last Minute Over Terrorist Threat', available at: www.guardian.co.uk/world/2008/jan/05/france.sport.
Hoffman, B., 'The Myth of Grass-Roots Terrorism', *Foreign Affairs*, May/June 2008, available at: www.foreignaffairs.com/articles/63408/bruce-hoffman/the-myth-of-grass-roots-terrorism.
Independent Monitoring Commission, Twenty-First report, 7 May 2009, available at: www.independentmonitoringcommission.org/documents/uploads/Twenty-First%20Report.pdf.
Independent Monitoring Commission, Twenty-Second report, 4 November 2009, available at: www.independentmonitoringcommission.org/documents/uploads/Twenty-Second%20Report.pdf.

International Olympic Committee, available at: www.olympic.org/en/content/The-IOC/.
MI5 website, available at: www.mi5.gov.uk/output/news/director-general-gives-interview-to-newspapers.html.
Olympic Charter, available at: www.olympic.org/Documents/olympic_charter_en.pdf.
RAND, *Aptitude for Destruction*, Volume II, available at: www.rand.org/pubs/monographs/2005/RAND_MG332.pdf.
Sturcke, James, Myers, Paul and Smith, David, 'Togo Footballers Were Attacked by Mistake, Angolan Rebels Say', 11 January 2010, available at: www.guardian.co.uk/world/2010/jan/11/two-arrested-togo-football-attack.

3 Al Qaeda and the London Olympics

Afzal Ashraf

Introduction

The 2012 London Olympics will provide two of the main ingredients that constitute an attractive terrorist target: a large concentration of people and ready availability of mass international media. Certainly, the global Jihadi terrorists, of whom Al Qaeda is the dominant, if not sole, brand, could be attracted to an event that amasses people from all over the world and provides the prospect of instant global publicity for any terrorist attack they seek to carry out. How likely that will be is impossible to fully predict. At this stage it is, however, possible to understand the factors that will determine any decision by terrorists to attempt an attack. These factors can be found through an analysis of past Olympic Games, from better understanding of Al Qaeda and from the lessons of recent counter-terrorism actions.

The threat posed by a terrorist group is a combination of its intent and its capability. This chapter will concentrate on factors that influence intent rather than methodology of attack or the capability of groups to carry out an attack. To determine if Al Qaeda will plan to attack the London Olympics, it is first necessary to understand Al Qaeda's ideological inheritance, as this is linked to what Al Qaeda wants to achieve. Then the role that terrorism plays in achieving its aims needs to be understood. It is also important to expose the peculiar nature of Al Qaeda, especially its influence on affiliated groups and individuals who share its aims so that their particular security challenges can be exposed. To begin with, the relationship between terrorism and the Olympics needs to be briefly explored to gain a historical perspective.

Words matter

Explanation of some definitions and terms is necessary at the outset. The Al Qaeda phenomenon has introduced new terminology into the Western lexicon. These terms are rarely defined and frequently misunderstood. Confusion or offence can sometimes result. 'Islamism' is a label used to describe movements in Islam that are primarily political in their outlook. The term 'Islamism' is therefore useful in differentiating between such movements and the purely theological Islam. For example, schools providing only education in Islam could be described as 'Islamic', but a school dedicated to a political movement in Islam could be described as an 'Islamist' school. Most Islamists do not support the use of violence to achieve their aims, but those that do are referred to as 'Islamist extremists'. Extremists who believe in the use of aggressive violence justify their stance using the Islamic concept of 'jihad', and so can be referred to as 'jihadi'. As will be shown later, most Muslims consider jihad to be a spiritual or defensive concept and so attachment of terrorism to a holy and noble term is uncomfortable for them.

Al Qaeda has been described as a 'Global Salafi Jihadi' movement because of its global aspirations and because its founder was brought up in the Salafi sect of Islam. Salafism is the predominant sect in Saudi Arabia and has its roots in what is sometimes called 'Wahabism', a puritanical and literalist movement started by Muhammad ibn Abd-al-Wahhab in the eighteenth century. However, the label 'Wahabism' is not accepted by most Salafis and also, they do not believe that bin Laden represents their view of jihad and so they object to the term 'Salafi' being attached to Al Qaeda.

Without clarifying these distinctions, it is easy to offend the sensibilities of Muslims, especially when referring to Islamist, Salafi or Jihadi terrorism. Terrorism is simply not allowed in Islam according to most Muslims and the apparent linking of Islamic terms to the word 'terrorism' is distressing for them. Unfortunately, the terrorists describe themselves as Muslims and their actions as being jihad in the service of Islam. In the absence of suitable alternatives, and because they are now in common usage, the terms will be used in this chapter, but with the caveat that they are not representative of majority or orthodox views. It would be helpful if those involved in Olympic security also developed a policy on terminology appropriate for security assessments and for public relations. Otherwise, confusion and unintended offence will remain a possibility.

Terrorism and the Olympics

The relationship between terrorism and the Olympic Games is not a new one. On the morning of 5 September 1972, the Palestinian terrorist group Black September seized 11 Israeli athletes and coaches in the Munich Olympic village. The ordeal ended with all 11 Israelis being killed, along with five of the eight terrorists. The event was broadcast as it unfolded throughout the world. Estimates suggest that over 500 million people witnessed the event on television, giving the terrorists unprecedented publicity.

The role of sport in modern society, the prestige associated with the Olympic Games, the huge commercial investment and the media involvement, all make the Olympic movement a significant actor in international relations. The Games themselves are a major international event. This symbolic relationship with the global interstate community means that the Olympic Games have been assessed by some analysts to be 'representative of the international status quo'. This in turn means that assaults on the Games 'whether through boycott, terror or propaganda, are possible because the Olympic system is an expression of the political status quo'.[1] The Munich Olympics marked a clear transition from terrorism being local or regional to it becoming global. This also marked the point when terrorists switched from attacking political targets to what are considered to be social targets. It was this transition, or the evolutionary nature of terrorist methodology and targeting, for which the Olympic security organisation at the time was totally unprepared.

The attack on the Munich Olympics was considered to have been a rational strategic choice to use terror. It was a form of compellence targeted at Israel, the international community, the Arab states and even the leadership of the PLO, who were considered by the Black September organisation to be insufficiently revolutionary and aggressive. Their objectives were to be achieved through terror amplified by the world's ever-present media. The media helped make it one of the most successful terrorist events in terms of publicity at the time, and it probably remained so until the events of 9/11. Most serious authorities on terrorism consider that terrorists are essentially realists who make rational choices in their use of violence. These choices are often a miscalculation in terms of effectiveness, but are nevertheless based on a belief that they will succeed in persuading their adversary to accept the terrorists' demands. Groups like Al Qaeda who invoke religion to justify their use of violence are considered by some to be irrational in their behaviour, but closer examination of their history, their objectives and their behaviour indicates otherwise.

Al Qaeda's ideological inheritance

In many ways, it is apt that no single event or date exists to mark the creation of Al Qaeda. Al Qaeda is the progressive evolution of a militant form of political Islam that began around the nineteenth century as a reaction to Western colonialism and cultural dominance. That ideology developed progressively in reaction to a number of Western interventions in Muslim countries, as well as the almost consistent failure of Muslim regimes to successfully enact modern forms of governance.

Islamism arose out of the belief that the loss of political independence and the cultural subjugation experienced by Muslims since the eighteenth century was due to their straying from the 'straight path' of Divine guidance. Whilst that belief along with the solution of a return to the pristine teachings of Islam is common to most strands of Islamic thinking, the Islamists further believe that purification of Muslim society requires a political and sometimes militant struggle called 'Jihad'. This concept of returning to a past, which is to some extent imagined and to a large extent unachievable in the modern world, allowed for an interpretation of Islam that owed more to the political and social aspirations of some ideologues than it did to scriptural and historic norms. It required a rejection of the prevailing international political, legal and economic systems as these were considered to be man-made orders and so a challenge to God's sovereignty on Earth. It required the need to establish alternative political, legal and economic structures based on God's law, the 'sharia'.

'Jihad', which means 'struggle' in Arabic, is used in two ways in Islamic theology. The primary meaning refers to the inevitable struggle between humans and their egos when they attempt to submit to the will of God rather than to their own selfish desires. This is known as the 'greater jihad' and is conducted through the affirmation of belief, through worship and through the exercise of morality. The other meaning, the 'lesser jihad', is fighting in defence of the freedom of belief or conscience.[2] The Quran has the following to say on this type of jihad:

> Permission to fight is given to those against whom war is made, because they have been wronged – and Allah indeed has power to help them – Those who have been driven out from their homes unjustly only because they said, 'Our Lord is Allah' – And if Allah did not repel some men by means of others, there would surely have been pulled down cloisters and churches and synagogues and mosques, wherein the name of Allah is oft commemorated. And Allah will surely help one who helps Him.[3]

The verse indicates that Allah is the same God who is worshiped by people in temples, churches, synagogues and mosques, and that all of these people have the right to defend their freedom of belief. As that belief is in an Almighty God, then that God has the power to guarantee success. This guarantee appeared to be fulfilled in early Islamic history when the Muslims were hugely outnumbered and outgunned but were victorious in the end. Their amazing victory was considered to be a sign of the truth of their message. This Divine guarantee of success is exploited by Al Qaeda to both confirm that it is 'rightly guided' and to reinforce its belief in an ultimate triumph.

Although Islam allows the use of force for just political and legal reasons, many classical Islamic scholars do not consider those activities as jihad in the above sense. It is only defending against persecution based purely on religious belief that strictly qualifies for the status of jihad in the above verse. The Islamists, however, conflate religion and politics and so make no such distinction. For Islamist extremists, almost any use of force by Muslims qualifies as jihad. Indeed, they consider fighting to be a form of devotion and refer to it as the sixth pillar of worship[4] for Muslims, and have made this form of jihad both a collective and an individual obligation. It is this obligation that motivates virtually all terrorist and most insurgent groups operating both in the West and against Western interests in Muslim countries.

The Islamists' concept of sharia raises a conundrum. On the one hand, it requires a cooperative approach to establish the conditions in society necessary to establish Divine law. On the other, it presents a problem, as there is only one God, there can be only one sharia. Islam, however, has a long tradition of diversity of belief and interpretation, making it impossible for all strands to agree a definitive statute of Islamic law. This conundrum remains unresolved. In the meantime, the jihadi Islamists have come to a realist accommodation. They have set aside their considerable theological differences and united around the idea of a violent jihad against the West. The Salafi-inspired Al Qaeda and the Deobandi

Taliban have found common cause in Pakistan, despite their significant differences over theology. The irony is that should they ever succeed in establishing a common homeland, they would be forced to wage war against each other in order to establish their particular versions of sharia.

In the meantime, those elements of Salafi, Deobandi, Berailvi and other Islamic sects who oppose violent jihadists are considered by the extremists to be disbelieving apostates (a concept known as *takfir*) who deserve to be killed. Just as the idea of jihad is used to justify the killing of non-Muslims, the idea of *takfir* is used to sanction the killing of Muslims by extremist Islamists. Using *takfir*, extremist ideologues have inspired the killing of thousands of innocent Muslim men, women and children – many times more than the number of non-Muslims killed through their particular version of jihad. A recent study by the West Point Centre for the Study of Terrorism found that over 85 per cent of Al Qaeda's victims were Muslims, a trend that is increasing.[5]

Justification for terrorism – religious or political?

Terrorism has been used by many groups throughout history and has been accepted as a 'legitimate' form of political violence by many. But few have felt the need to provide elaborate religious justifications for their acts. Terrorism by Jewish groups such as the Irgun and the Stern Gang, in support of the creation of Israel, did not rely on any significant scriptural justification from the Old Testament, although many members were inspired by the Zionist myth of salvation through return to the state of Israel.[6] The Palestinian Liberation Organisation's terrorism in the 1970s and 1980s was conducted under the Marxist ideology of revolutionary movements that accepted violence as an essential part of political change. Even the UK's Foreign Secretary, David Milliband, postulated in 2009 that terrorism was a justifiable necessity in certain circumstances.[7]

So why is there such a strong religious dimension to Al Qaeda's terrorism? The peculiar nature of Islamist terrorism is that it claims to be acting in the name of Islam. But of all the religions, Islam has the most explicit prohibitions against the use of force, especially against non-combatants. Anyone who claims to kill non-combatants has, therefore, to provide a compelling theological justification for their actions. The evolution of Islamist thought has allowed Al Qaeda to adapt – or pervert, as some suggest – the Islamic concepts of *takfir* and Jihad for violence against Muslims and non-Muslims. It is worth noting that Hezbollah in Lebanon and Hamas in Palestine also use similar jihad and retribution arguments to justify the deaths of Israeli civilians and children. But their justification narratives tend not to encompass Israeli and Western targets outside their regions. Also, these organisations place little emphasis on *takfir* and so rarely target other Muslims on a sectarian basis.

Al Qaeda, however, has developed strong rhetorical justifications for attacking non-combatant men, women and children. It has progressively expanded the envelope of acceptable targets. Before 9/11, bin Laden berated the West for killing women and children in its military operations, boasting that the Mujahideen had not done so during the war against the Soviets in Afghanistan:

> I [do not] consider the killing of innocent women, children and other humans as an appreciable act. Islam strictly forbids causing harm to innocent women, children and other people. Such a practice is forbidden even in the course of a battle. It is the United States, which is perpetrating every maltreatment on women, children and common people of other faiths, particularly the followers of Islam.[8]

But the difficulty of attacking legitimate Western targets such as the military forced Al Qaeda to employ mass-terror tactics. After 9/11, bin Laden was forced to defend the killing of civilians by declaring that as the USA killed Muslim civilians and as it was a democracy, US citizens were vicariously guilty of their government's action and so deserved to be killed:

> The American people should remember that they pay taxes to their government, they elect their president, their government manufactures arms and gives them to Israel and Israel uses them to massacre Palestinians. The American Congress endorses all government measures and this proves that the entire America is responsible for the atrocities perpetrated against Muslims. The entire America, because they elect the Congress [sic].[9]

Another argument used by him to defend the killing of innocents is that of just retribution: the killing of Western civilians is just retribution for the killing of Muslim civilians by Israel and Western governments. Non-Western civilian casualties caught up in an Al Qaeda terrorist attack are dismissed on the grounds that they will probably come from a country that is guilty of supporting or failing to stop the West in its actions against Muslims or as merely collateral damage. Al Qaeda's promiscuity for violence is such that now virtually any target and any method of attack is justifiable according to its ideology. That ideology may be presented in religious terms but its aims and arguments are based on the rational of Realpolitik. It is Realpolitik that drives Al Qaeda's strategy of terrorism.

What does Al Qaeda want?

If there is a point at which Al Qaeda came into existence, it was about three years after the first Gulf War, in the mid-1990s. When Saddam invaded Kuwait in 1991, bin Laden offered the Saudi king his services to evict Saddam, arguing that, as he and his mujahideen band of brothers had successfully defeated a superpower in Afghanistan, Saddam would pose no great challenge. He warned against allowing the West to be involved because the West would use the situation to occupy Muslim lands and to take Muslim wealth. Three years after the liberation of Kuwait, Western forces remained stationed in Saudi Arabia and the country went through a virtual bankruptcy. In the words of the US Secretary of State, Baker, 'The Saudis paid for the war, and some!' The resulting political concerns over perceived US 'occupation' and economic hardship gave rise to dissent, which the Saudi government suppressed, increasing the perception of their being American stooges.

The situation confirmed in bin Laden's mind his concerns that the US was leading a Western war against Islam and that most Muslim governments were rendered impotent in enforcing the interests of Muslims through a combination of corruption, incompetence and Western manipulation. Bin Laden's success in predicting the situation and warning the Saudi king several years previously further confirmed to him that he had a better understanding of international politics and how to protect Muslim interests than the leaders of the Muslim world. This 'superior wisdom' had come to him through his knowledge of Islamist ideology and through his experience of the Jihad in Afghanistan in the 1980s. It was a jihad in which he would have believed that the Qur'anic promise of success against all odds was spectacularly fulfilled when a rag-tag guerrilla army caused the downfall of a superpower. In the meantime, he had left his homeland for the Sudan where he assuaged his desire to serve his faith by building roads and other projects. The Sudanese government, under pressure from the West, invited bin Laden to leave despite his considerable personal investment in developing that country. Bin Laden returned to Afghanistan where the newly established Islamist Taliban government welcomed him.

The entire episode resulted in bin Laden making his most significant contribution to the evolution of Islamist ideology; the concept of the near and far enemies. Bin Laden reasoned that up to that point, all Islamist-inspired rebellions had concentrated on the apostate Muslim regimes in their locality. They had failed because these regimes were underpinned by Western support. Success lay in forcing the West, the far enemy, to give up on the idea of choosing and supporting regimes in the Muslim world so that Islamists could move in to take power. Concentrating on attacking the far enemy was therefore to be the cornerstone of the new Islamist movement that was to become known as Al Qaeda and it would of necessity be a global movement.

The objectives of Al Qaeda were articulated as: the driving out of Western 'occupation' forces from Muslim lands and the overthrow of 'apostate' Muslim regimes. The resulting political vacuum was to be filled by the establishment of Islamist states governed according to its interpretation of sharia under an international caliphate.

How does Al Qaeda use terrorism?

Al Qaeda's strategy for achieving its objectives before 9/11 was to use 'shock and awe' mass-casualty attacks against US military and political targets. Bin Laden reasoned that the US's response to Vietnam in the 1960s, the bombing of the USMC barracks in Lebanon in 1983[10] and the Somalia episode in 1992[11] showed it had little appetite for absorbing even relatively small numbers of casualties because the USA withdrew its forces in each instance. Al Qaeda consequently embarked on a series of high-profile attacks beginning with the Twin Embassy bombings in Tanzania and Kenya in 1998,[12] the attack on USS *Cole* in 2000[13] and culminating with the 9/11 attacks in the USA.

None of these attacks persuaded the USA to withdraw, and the 9/11 attack in particular increased its resolve to maintain a presence in the Middle East. Bin Laden had failed to appreciate the difference between what the US considers 'wars of choice' and 'wars of necessity'. He ought to have used Pearl Harbor as a model for the USA's response to 9/11 rather than the conflicts in Vietnam or Somalia. The USA has a different approach to taking casualties when it believes its territory is under attack. Under these circumstances, it considers it necessary to fight with a determination that is absent in conflicts that it chooses to embark on for a foreign-policy objective. Rather than retreating from the Middle East, the USA embarked on a 'War on Terror' that resulted in unprecedented political and military intrusion into Muslim lands with consequentially high casualties amongst Muslim civilians as well as its own forces.

Al Qaeda rapidly adapted to the new situation by seeing the Coalition attack on Afghanistan in 2001 and the invasion of Iraq in 2003 as opportunities for insurgency warfare. Bin Laden and his cohort had more experience of insurgency warfare than of terrorism, given the decade they had spent fighting the Soviets in Afghanistan. Not only had they gained training in operational planning and tactical know-how, but also they had gained something more powerful – the confidence of winning against a superpower. Bin Laden devised a new strategy of 'spreading' or depleting the West politically, militarily and economically.

This strategy required the frustration of the West's plans in Muslim countries, particularly in Iraq and Afghanistan, to deplete its political capital. It required attritional engagement with Coalition militaries to destroy their aura of invincibility and it required a prolonged campaign in order to escalate the costs of the war and to prevent any recouping of losses through investment in the reconstructed economies of the countries involved. All of these objectives would be achieved through an insurgency campaign in which terrorism played a supporting role rather than being the primary tactic in what Al Qaeda referred to as 'occupied lands'.

A similar shift of emphasis also occurred in Al Qaeda's approach to attacks on Western soil. The failure of terrorist attacks to compel the West to withdraw from Muslim lands, and the opportunity of insurgency-based warfare to defeat the West, resulted in terrorism being relegated from a strategic to a tactical role. The most significant Al Qaeda attack in the UK after 9/11 was the London Bombings of 7/7 in 2005. The leader of that attack, Sadiq Khan, stated in his martyrdom video that the London attacks were a reprisal for Western atrocities against Muslims and that such attacks will continue as long as the West continued to conduct operations against Muslim lands.

That message indicated that Al Qaeda no longer expected to deliver a knockout blow to the West through terrorist acts as it had hoped to do before 9/11.

Rather, terrorist operations on Western soil were intended as symbolic acts to create a sense of insecurity in the West in return for the perceived insecurity created by the West in Muslim lands. They are also intended to demonstrate the failure of the West to achieve success in its War on Terror, which Al Qaeda alleges is in fact a war on Islam. Whilst these two imperatives make Al Qaeda-inspired terrorism in the West important to it, the objectives of that terrorism are more symbolic than strategic and so terrorist operations attract lesser priority than the insurgency-led campaigns in Iraq, Afghanistan and Somalia, for example. These priorities are likely to remain unchanged as long as Al Qaeda feels that the insurgencies in Afghanistan and Pakistan are achieving their aims. Should there be a radical change before the London Olympics, such as the cessation of hostilities, then Al Qaeda may revert to prioritising spectacular acts of terrorism on Western soil.

This is not to say that Al Qaeda is investing little effort in direct attacks in the West. On the contrary, it remains keen on mass casualty attacks in the West, especially in the UK and the USA given their sizeable involvements in Iraq and Afghanistan. The fertiliser bomb plot[14] and the airline plot[15] are just two of the potentially devastating examples of terrorist operations that have been foiled by security services in the UK. To what extent these attacks are linked to Al Qaeda is debatable, as Al Qaeda is not a conventional organisation with a recognised membership and organisational structure.

Challenges posed by Al Qaeda, its associates and the 'inspired'

Al Qaeda core

Al Qaeda is a movement that has at its core an idea and a leadership of ideologues. Osama bin Laden and Ayman Zawahiri are the two principle ideologues. Bin Laden is certainly the one that most personifies the character of Al Qaeda and is by far the most charismatic leader in the movement. This core leadership structure includes a number of committees, or *shuras*, dealing with operational, political, media, religious and other issues. The terrorist attacks mentioned so far have been, to varying degrees, instigated, planned, supported and conducted by individuals from within this Core structure. That would invariably have involved direct contact between at least one member of the terrorist cell and Al Qaeda Core. For that to happen, the terrorists would have needed to travel to Pakistan or Afghanistan to receive instructions, training and, above all, motivation.

Al Qaeda associates

There is another layer of organisations called 'Al Qaeda Associates'. These include groups such as Al Qaeda in Iraq, Al Qaeda in the Islamic Maghreb, Al Qaeda in the Arabian Peninsula, etc. These groups work virtually autonomously and concentrate their efforts against local or regional targets (the near enemy) but they will, if the opportunity presents itself, attack Western (far enemy) targets. The June 2007 failed suicide attack on Glasgow airport has been linked to the Associate group Al Qaeda in Iraq.[16]

Al Qaeda in the Islamic Maghreb, for example, may have an aspiration to conduct attacks in Europe given its proximity to Spain and Italy. Maghreb Diaspora communities such as Moroccans in the Netherlands and Algerians in France could provide cover for Al Qaeda operatives from the Maghreb group.[17]

Al Qaeda-inspired groups and individuals

The most successful post-9/11 attack in terms of casualties and in influencing Western political policy was that carried out against the Madrid train network on 11 March 2004.[18] The attack precipitated the Spanish withdrawal from the Iraq coalition. The decision to withdraw was not in direct response to the terrorist attack but came about because the incumbent government, which was committed to staying in Iraq, made a political miscalculation. It tried to blame the attacks on the Spanish terrorist group ETA. The opposition party, who were committed to withdrawal from Iraq, exploited a public feeling of being misled on the cause of the bombings. That party won the general election just a few days later and fulfilled its promise of withdrawal. The incident illustrates that terrorists can achieve their political aims through a flawed reaction to an act, as much as they can through the act itself. A group that appears to have had no direct link with Al Qaeda carried out the Madrid bombings. Such groups are known as 'Global Jihadi inspired groups'.

There are examples of Al Qaeda inspired individuals who have attempted to commit terrorism on British soil. Andrew Ibrahim, a young convert to Islam, made explosives and a suicide vest for an attack on a Bristol shopping centre, in 2008. Nicky Reilly, another convert, partially blew himself up in an Exeter restaurant in the same year. The fact that both individuals were converts highlights another relevant aspect of Islamist extremist terrorism in Britain. A disproportionately high number of terrorists are converts. Other converts include: Richard Reid, the shoe bomber; Jermaine Lindsay, who was one of the 7/7 bombers; Simon Keeler, a 36-year-old member of al-Muhajiroon who was convicted of inciting terrorism in March 2008; and two of the airline bomb plot suspects. Many of them had only recently converted and their conversion was not so much to the religion of Islam but to the ideology of Islamist extremist violence. Such converts appear to select militant forms of Islam for what Olivier Roy call a 'protest identity';[19] extremist Islamism provides them with an ideological umbrella to violently respond to a range of grievances they have against Western society. Much of the British government's Prevent initiatives are targeted at established Muslim communities. Converts present a particular problem as they are generally either outside these communities or have only peripheral contact with them.

There is a view among terrorism experts that 'inspired' groups and individuals represent a lesser threat than those who are supported by Al Qaeda Core because of the superior training and planning skills that Al Qaeda can provide. That may be true as a minor statistical trend, but as the Madrid example shows, a devastating attack by non-Al Qaeda Core supported or Affiliated members is very possible.

The probability that future attacks will come from such individuals or groups is increasing due to constraints placed on Al Qaeda by the recent spate of airstrikes in Pakistan's tribal region and in Afghanistan. The airstrikes have reduced the ability of individuals to travel to and from these regions as well as reducing the ability of Al Qaeda to host training on the same scale as it used to. These factors could encourage those who would otherwise be tempted to go abroad for training to stay at home and plan, prepare and execute attacks using Internet-based research. They form part of what Marc Sageman calls the 'leaderless jihad'.[20]

These 'inspired' individuals or groups are hampered by the fact that travel routes to places of concern such as Pakistan and communications to known jihadi extremists who facilitate operational support or training are being internationally monitored. Inspired groups and individuals are unlikely to use these conduits to get in touch with Al Qaeda and so reduce their chances of being detected. This appeared to be the case with Nicky Reilly, who had not travelled to another country to be radicalised by an extremist group or to learn terrorist tradecraft. He only came to notice when his homemade bomb partially detonated in the toilet of the restaurant he had intended to blow up. A tip-off to Special Branch from the Bristol Muslim community thwarted Andrew Ibrahim's plan, possibly just days before he intended to execute his deadly act. Incompetence on the part of the terrorists and lucky tip-offs for the police helped prevent these cases from causing loss of innocent lives, but they cannot be relied on as part of a security strategy to safeguard the London Olympics.

Al Qaeda and London 2012

Lessons of post-9/11 Olympics

The two post-9/11 Olympics – Athens in 2004 and Beijing in 2008 – had the spectre of an Al Qaeda attack hanging over them. In neither case was any significant plot discovered, let alone a terrorist attack carried out. In fact, there have only been two successful terrorist attacks against the Olympic Games, and neither of them has been inspired by an Islamist organisation. That could be taken as reassurance that Olympic security has a good track record in recent years and that Global Jihadi terrorists are not too interested in attacking the Games.

However, neither China nor Greece represented as attractive a target to Al Qaeda as the UK does with its past involvement in Iraq and its current involvement in Afghanistan. The UK is also significant because of its wider economic and political role in the Middle East, especially because of its historic relationship with Israel. China, on the other hand, helped the mujahideen during the

1980s Afghan campaign. It also allegedly had relationships with the Taliban government and with Al Qaeda,[21] making it a less attractive target. Furthermore, Greece did not have a suitable community within which Al Qaeda sympathisers could easily exist. Although China has a large Muslim Uighur population, many of whom have political and economic grievances against the government, the State has for a long time kept close control over external visitors and influences in the region. This would have made it difficult for Al Qaeda to establish a bespoke attack cell in China. Sending external operatives to both these countries sufficiently early to conduct the necessary surveillance and planning for an attack without arousing suspicion could have been highly risky and problematic for Al Qaeda.

Previous Olympics do not, therefore, provide a good indicator for the threat to the London Olympics. Furthermore, attacking sporting events has subsequently become part of the Global Jihadi doctrine. The *Encyclopaedia of Afghan Jihad* has for many years included football stadiums amongst its list of suitable targets.[22] The 2009 Lahore attack on the Sri Lankan cricket team during its tour of Pakistan shows that sporting targets have moved from the realm of the theoretically possible to the realm of practical reality.

Decision to act

The point is therefore not whether Al Qaeda will want to attack the London Olympic Games but whether it is willing and able to commit the resources necessary to plan and mount an attack that is both spectacular and successful – two of the criteria that apparently drive Al Qaeda Core's decision-making process. Spectacular attacks and success are important to Al Qaeda's image of a highly competent and deadly organisation able to strike in a manner that shocks and is awe-inspiring. That is why meticulous planning and training are very much part of its modus operandi, as has been witnessed by the twin Embassy bombings in East Africa, the attack on the USS *Cole* and, of course, 9/11. The Olympics pose both opportunities and challenges that are significantly different from Al Qaeda's usual modus operandi of attacking established facilities or routine transport services.

Al Qaeda Core's modus operandi requires that they, not the West, choose the time, place and method of attack, so as to maximise the shock value of the attack and to minimise the possibility of detection or disruption by security forces. Attacking the Olympics would be a significant achievement but it would not be entirely unexpected. The heightened security presence around the Games increases the risk of detection and disruption of any attack. In the past, terrorists have chosen to mitigate the risk by conducting attacks in areas away from Olympic venues where security would have been lower. An example of this is the attack in China's Kuming city just a few days before the 2008 Olympics.[23] Coordinated explosions on buses killed at least three people and were claimed by the Turkish Islamic Party. The attacks were hardly reported and the group was unable to exploit them for any significant political purposes.

The counter-terrorism campaigns conducted in Afghanistan, Pakistan and in other regions have led to the emergence of new affiliates who wish to join Al Qaeda. These include the al-Shabaab in Somalia and Indonesia's Jamaah Islamiah splinter group, Tadzin Al Qaedat. Should the number of these groups, their capabilities and their global aspirations continue to grow, then they will considerably complicate the security situation and the intelligence and prevention challenges will also consequently grow.

Community engagement

Detecting and monitoring extremists is a recognised challenge. The site of the London Olympics happens to be surrounded by a number of Muslim communities within which potential terrorists or radicalisers may lurk.[24] The sincere desire for a safe and peaceful existence that exists within the vast majority of ordinary Muslims in these communities ought to be harnessed and they should be helped to detect any signs of suspicious behaviour and feel empowered to act.

Compelling evidence suggests that good police and community relations can prevent serious terrorist plots. The fertiliser bomb plot was foiled when staff at a self-storage unit in London reported their suspicions to the police. Andrew Ibrahim's plot was disrupted when a member of the Bristol Muslim community became suspicious of Ibrahim's statements and behaviour. The local Special Branch had previously established good relationships with the Muslim community and had left the contact details of its liaison officer with them. The Special Branch liaison officer was actually on holiday at the time when he received the tip-off. But because his personal mobile number was with community members, the concerned Muslim citizen was able to speak with him. The liaison officer immediately passed the tip-off to the police's counter-terrorism squad who apprehended Ibrahim and discovered explosive material, suicide vests and other apparent indicators of an imminent terror attack. These incidents illustrate the level of trust and personal access between the police and communities that is crucial to successful counter-terrorism.

Other instances have caused police and community relations to suffer as a result of counter-terrorism actions. A Muslim man was shot and wounded during a police raid on his house in Forest Gate.[25] The raid resulted in no charges against the occupants of the house. Another series of raids in Liverpool and Manchester in April 2009, in which 12 men (including ten Pakistani students) were arrested in a high-profile operation, also failed to yield any charges of terrorism, although some of the arrested individuals were nominated for deportation on other charges. Such events have led to a loss of confidence from sections of Muslim communities who believe that the police are unfairly targeting them. The police would argue that they had intelligence in both cases that required them to swiftly act in order to prevent possible danger to the public.

To gather sufficient information to convince a jury beyond reasonable doubt that a crime has been committed requires time for a meticulous investigation. The police, however, would find it difficult to justify delaying disruptive action when they are in possession of intelligence that raises even the slightest suspicion of a risk to public safety. Consequently, counter-terrorism policing will inevitably result in instances where the police will take action that results in no charges or conviction.[26] This imbalance between the veracity of information necessary to force the police to take disruptive action in the interest of public safety and that needed to form evidence for a successful criminal prosecution has not been fully explained to the public. British Muslim communities certainly need to understand this subtle issue so that they can avoid interpreting police actions, the majority of which will not lead to prosecutions, as harassment. Otherwise the success of cases like Andrew Ibrahim's may not be repeated.

Internet and ideologues

The Ibrahim and Reilly cases highlighted the role of the Internet and ideologues in radicalising the individuals. It has been reported that Nicky Reilly was ideologically encouraged by a Pakistan-based Internet contact.[27] Virtually every terrorist act, especially religiously justified terrorism, is inspired by an ideologue. Abu Hamza[28] and Abu Qatada[29] are two extremist ideologues whose preaching has influenced young British men and whose ideas are accessible from the Internet. These ideologues take past and current world events to construct a singular narrative that depicts a conspiratorial conflict between the West and Islam. Innocent Muslims are politically, physically, economically and culturally violated in this conflict. They are helpless because Muslim governments have been politically neutered by Western powers. Whilst these narratives use historic events like the Crusades and the creation of Israel, they increasingly rely on the immediacy and vivid imagery of contemporary grievances, usually provided by the Internet. Abu Graib, Guantanamo Bay and the 2008 Israeli bombing of Gaza are just some examples of events that ideologues exploit to create powerful feelings of indignation, crisis and the necessity for an urgent and violent jihad. These narratives and calls for action are relayed through the Internet in the form of speeches, sermons and papers. Other Internet sites provide information on how to channel the anger of jihadi disciples by giving information on surveillance, weapons and bomb-making. As long as ideologues have access to the Internet, they will get their message out and are likely to be able to influence vulnerable individuals.[30]

Conclusions

History affirms: 'successful counter-terrorism campaigns have "outthought" rather than "out-fought" the terrorists.'[31] Even terrorists acknowledge: 'Terrorism is a thinking man's game.'[32] Counter-terrorism therefore requires education in all aspects of the threat. This will be especially necessary for the surge in the number of individuals drafted in to deal with Olympic security.

Al Qaeda shares an ideological inheritance with modern Islamists who wish to establish Islam as a political force in the world. It has taken their theological and political ideas and converted them into a strategy based on violent conflict to achieve a number of specified aims. Its justifications for violence against non-combatants are expressed in religious terms but are in fact secular expressions of Realpolitik. The result is that it now has no restraints on who it will target and how it will do so.

Olympic Games have been an attractive iconic target for terrorists in the past, but Al Qaeda has made no serious attempt to target previous Games. That does not necessarily mean it will not attempt to do so this time. Acts of terrorism on Western soil are an important priority for Al Qaeda but they are subordinate to the current insurgency campaigns it is supporting in Afghanistan, Pakistan and elsewhere. Al Qaeda's current condition in the tribal areas of Pakistan is such that it will find it difficult to mastermind any complex attack against the inevitably high security at the Olympics.

A growing number of Associates, such as Al Qaeda in Iraq and Al Qaeda in the Islamic Maghreb have complicated the analysis of Al Qaeda's intentions. Al Qaeda-inspired individuals and cells present a particular challenge. Reacting to these and any emerging security challenges to the Olympics will be best done by responsible senior officials having as good an understanding of Al Qaeda's aims, ideology and influence as possible. An appreciation of the role of ideologues, especially their recruitment and exploitation of converts, will be useful. Community education and awareness, particularly amongst Muslim communities, will be an essential component of any counter-terrorism strategy. Good police and community relations have in the past thwarted terrorist attacks and remain one of the most effective responses available to civil society.

Notes

1 David B. Kanin (1981) *A Political History of the Olympic Games*. Boulder: Westview Press, p. 6.
2 As with most religious concepts, there are differing interpretations. Some Muslims, especially most militant Islamists, do not recognise the concept of a spiritual Jihad and only interpret Jihad in a military sense.
3 Qur'an, Chapter 22: Verses 40–42.
4 Most Muslims recognise the declaration of faith, prayers, observing the fast in the month of Ramadan, performing the Hajj (if possible) and paying Zakat (alms for the needy and poor) as the five pillars of worship.
5 Scott Helfstein *et al.* (December 2009) *Deadly Vanguards: a Study of al-Qa'ida's Violence Against Muslims*. West Point: Combating Terrorism Center at West Point. The Report found that:

> From 2004 to 2008, only 15% percent of the 3,010 victims were Western. During the most recent period studied the numbers skew even further. From 2006 to 2008, only 2% (12 of 661 victims) are from the West, and the remaining 98% are inhabitants of countries with Muslim majorities.

6 See R.C. Rowland (1985) *The Rhetoric of Menachem Begin – the Myth of Redemption Through Return*. Maryland: University press of America.
7 Speaking on BBC Radio 4 on the *Great Lives* programme (Series 19 on 11 August 2009) on the topic of struggle against the South African apartheid regime.

8 Osama bin Laden in an interview with the Pakistani newspaper *Ummat*. Published 28 September 2001.
9 Bin Laden in interview with Hamid Mir of Pakistan's *Dawn* newspaper on 21 November 2001.
10 On 23 October 1983, buildings housing US and members of the Multinational Force in Beirut were struck by trucks driven by suicide bombers. The 241 deaths, along with 60 injured, represented the highest single-day death toll for US servicemen overseas since the Second World War. The incident led to the withdrawal of the international peacekeeping force from Lebanon, where it had been stationed following the Israeli invasion of Lebanon in 1982.
11 US troops participating in a UN peacekeeping mission in Somalia in 1992 became trapped in Mogadishu. A total of 18 American soldiers were killed, and a television crew filmed the body of a US soldier being dragged through the streets by a mob. US forces were swiftly withdrawn following the incident.
12 On 7 August, suicide bombers using trucks laden with explosives simultaneously struck the embassies in Dar es Salaam and Nairobi. In Nairobi, approximately 212 people were killed, and about 4,000 were injured; in Dar es Salaam, at least 11 were killed and 85 wounded. Although the attacks were directed at the USA, the majority of casualties were locals, and only 12 American citizens were killed.
13 The USS *Cole* suffered a suicide attack on 12 October 2000 while it was harboured in the Yemeni port of Aden. In total, 17 US sailors were killed and 39 were injured.
14 Five members of a British Al Qaeda cell were jailed for life on 30 April 2007 for plotting to use fertiliser bombs to blow up the Bluewater shopping centre, the Ministry of Sound nightclub and other UK targets including gas and electricity supplies. They were arrested in March 2004 as part of security operation Crevice.
15 An alleged plot to blow up airliners between the UK and the USA and Canada using home-made liquid explosives smuggled onto scheduled flights in soft-drink bottles. In August 2006, eight men were charged as part of police operation Overt and only three were found guilty of conspiracy to murder in September 2008.
16 S. O'Neill, S. Bird and M. Evans (17 December 2008). 'Glasgow Bomber Bilal Abdulla was in Iraq Terrorist Cell', *The Times*. Bilal Abdullah, the only member of the gang to survive and be convicted, is of Iraqi descent and qualified as a doctor in Iraq. The report cites US sources to link him with Al Qaeda. Significantly, Bilal claimed through his QC that it was anger at the 'illegal war' in Iraq, not religion, that motivated his attack.
17 The ability of this group to carry out attacks in Europe is disputed amongst analysts. Some regard it as being incapable in Europe, while others disagree, pointing to reports that

> In June 2008, Spanish authorities uncovered a terrorist cell in Spain, arresting eight men and detaining ten accused of providing logistical and financial support to AQIM. This follows French police uncovering a similar cell in the outskirts of Paris in December 2007. Arrests of suspected terrorists with ties to AQIM have been made throughout Europe in the United Kingdom, Germany, Italy, Portugal, and the Netherlands.

See A. Hansen and L. Vriens (21 July 2009) 'Al Qaeda in the Islamic Maghreb (AQIM) or L'Organisation Al-Qaïda au Maghreb Islamique (Formerly Salafist Group for Preaching and Combat or Groupe Salafiste pour la Prédication et le Combat)', *Council on Foreign Relations*.
18 A total of 191 civilians died with over 2,000 injured.
19 See M. Uhlmann (2008) 'European Converts to Terrorism', *Middle East Quarterly*, XV (3), 31–37.
20 See M. Sageman (2008) *Leaderless Jihad Terror Networks in the Twenty-First Century*. Philadelphia: University of Pennsylvania Press. Bruce Hoffman disagrees

with Sageman's analysis, arguing that Al Qaeda is still able to centrally inspire terror attacks and that 'grassroots' terrorism is a myth. Fundamentally, both analysts are right. Their mistake is to regard each other's position as mutually exclusive. For Hoffman's critique of Sageman see B. Hoffman (2008) 'The Myth of Grass-Roots Terrorism', *Foreign Affairs*, 87 (3).
21 Roger Faligot (2008) makes these claims in his book, *The Chinese Secret Service: from Mao to the Olympic Games*. Nouveau Monde Publishers, Paris.
22 This manual contains volumes on military and terrorist related topics such as assassination and bomb construction. The publication is widely referenced on the Internet and has been described as a 'blueprint for terrorists'. In May 2004 its discovery at Abu Hamza's address provided the basis for one of the charges for which he was convicted in February 2006.
23 See www.telegraph.co.uk/news/worldnews/asia/china, 21 July 2008.
24 For more information on radicalisation and associated communities, organisations and locations, see R.B. Birdwell (May 2009) *Radicalisation Among Muslims in the UK – Policy Working Paper*. Policy Working Paper. Brighton: MICROCON.
25 Mohammed Abdulkahar, 23, was accidentally shot at his home in Forest Gate on 2 June by police conducting a raid in response to intelligence reports indicating the presence of chemicals for terrorist purposes in the house. None were found. He was later arrested on suspicion of making pornographic pictures of children but again no charges were brought.
26 D.A. Clarke (24 April 2007) 'Learning From Experience – Counter Terrorism in the UK Since 9/11', *The Colin Cramphorn Memorial Lecture* (p. 15). London: Policy Exchange. Deputy Assistant Commissioner Peter Clarke highlighted this conundrum during the lecture, but it did not appear to spark sufficient public debate for the issue to be widely appreciated and so supported by the public.
27 See A. Fresco, 'Bomber Nicky Reilly was brainwashed online by Pakistani extremists'. www.timesonline.co.uk, 16 October 2008.
28 Abu Hamza al-Masri took part in the 'jihad' in Afghanistan and Bosnia. He delivered many sermons and pamphlets encouraging hatred against Muslims, Jews and Westerners as well as supporting terrorism. He was jailed in February 2006 for 11 charges, including inciting hatred and murder.
29 Abu Qatada is alleged to be one of the most influential ideologues supporting Al Qaeda in Europe. He is alleged to have influenced terrorists. Videos of his sermons were reportedly found in the apartment of 9/11 ringleader Mohammed Ata (see, for example, 'Abu Qatada Gets £2,500 Compensation for Breach of Human Rights', www.guardian.co.uk, 19 February 2009).
30 For a more extensive but clear amplification of this issue see Joseph Lieberman *et al.* (May 2008) *Violent Islamist Extremism, the Internet, and the Homegrown Terrorist Threat*. United States Senate Committee on Homeland Security and Governmental Affairs. Washington: United States Senate.
31 G. Woo (n.d.) *A Terrorism Risk Analyst's Perspective On TRIA*. Risk Management Solutions. London, www.pcraroc.com/Publications/RiskAnalystPersectiveTRIA_ WooForCongress.pdf.
32 Dr George Habash, co-founder of the Popular Front for the Liberation of Palestine. Quoted in Woo (n.d.).

4 Understanding terrorist target selection

Andrew Silke

Introduction

Virtually every summer Olympics since 1972 has either experienced terrorist attacks or else has been the target for terrorist plots. The terrorist groups behind these plots and attacks have come from a wide range of backgrounds, with different ideologies, different aims and different histories. As already highlighted in Chapter 1 in this volume, the Olympics have been targeted by a range of nationalist separatist groups, Christian fundamentalists, Islamist extremists, militant Marxists and others. The organisations themselves have varied from fringe groups with only a handful of people involved to movements with hundreds if not thousands of active members and budgets which can run into tens of millions of pounds. States, too, have not been above targeting the Games for violence.

The one thing that all of these different groups have in common is that they judged the Olympics to be an attractive and appropriate target for violence. Despite their different ideologies and different objectives, all of these groups made a decision that the Olympics was the right target for them. It would be naive in the extreme to assume that no group will be interested in attacking future Olympics – or other major sporting events – when so many groups have shown both the willingness and the capability to do just that in the past 40 years.

Why such a wide variety of groups continuously target the Olympics is one of the focuses of this chapter. It examines terrorist target selection in detail, shedding light on how and why terrorist groups select some targets and ignore others. It will also explore how states attempt to block and deter terrorist attacks and how the terrorist groups themselves react to these efforts.

The calculus of terrorist targeting

Terrorist violence is often portrayed in the media as mindless and indiscriminate, but in reality terrorist attacks are usually the result of deliberate and considered planning, and the weighing up of different choices and options. Some terrorist attacks take years to plan – others just days or even hours (Gilmour, 1998). Attacks against the Olympics have been a mix, some plots involved long-term planning while others – including both Munich and Atlanta – were pulled together in a matter of weeks. A key to understanding terrorist target selection is to gain an insight into the decision-making that takes place during this planning stage.

One of the approaches that emerged to meet such needs is what is now known as the 'rational choice perspective'. Emerging originally from the field of economics, this approach argues that it is often more useful (and more accurate) to view criminal activity as having an underlying rational basis. Such a perspective assumes first that crime is a 'purposive behaviour which involves the making of (sometimes quite rudimentary) decisions and choices, which in turn are constrained by limits of time and ability and the availability of relevant information' (Clarke and Cornish, 1985: 163). There is an emphasis on focusing on specific crime episodes, as each type of crime will have its own particular chain of decision-making leading up to, and including, the commission of that crime. Several authors have already argued that such an approach can be a highly useful lens through which to understand terrorist incidents (e.g. Taylor, 1988; Jacobs, 1998; Silke, 2003).

The rational choice approach is certainly readily applicable to terrorism, though the factors affecting terrorist decision-making can be substantially different from those behind a financial criminal such as a burglary. The factors that lead to the undertaking of the terrorist act will certainly gain their meaning by reference to an ideology, but the choice of a particular activity (such as a bombing or shooting) and the circumstances in which it might occur, are affected by much more local issues.

Experienced terrorism researchers, such as psychologists Martha Crenshaw and Max Taylor and economists such as Todd Sandler and Walter Enders, have long argued that it is gravely mistaken to assume that rational processes do not underlie terrorist decision-making (and have a substantial body of research to support such a view) (e.g., see Silke, 2001). The rational choice approach, however, accepts that decisions can be made quickly and poorly, and are always subject to deficiencies in the available information and the abilities of the decision-maker. Thus, the rational choice approach is not a synonym for saying that terrorists always make good decisions. Rather, the approach emphasises that decisions (however rudimentary and quick) have always been made and that a range of factors must have played a role in the selection of one choice over another. *But which factors?* That is the question lying at the heart of the rational choice approach and once one can identify these critical variables, clear avenues to prevent and counter future acts of terrorism are opened.

The problem, though, for many observers is that often the perspective and inside knowledge needed to make this clear is lacking, and with extreme violence it is easier to assume there is no logic or rationality behind the violence, only chaos and insensible hate. However, in looking at a terrorist act from the cooler perspective of a rational actor, it can be possible to gain a far greater understanding of the process that leads to violent attacks than if the focus is simply caught up with the horror of the incident and speculation as to the type of person who would be capable of carrying out such atrocity.

This is not to say – or argue – that a rational choice perspective is the best or only way to study terrorism. Far from it. The point is merely that it can often be a highly appropriate lens through which useful insights and understanding can be made. Used in balanced conjunction with other perspectives, this approach can lead to a fuller grasp of the processes driving terrorism. As a framework, it arguably has particular value in contemplating the more horrific terrorist atrocities. The death and suffering of large numbers of innocent men, women and children is profoundly emotive for any observer – even one charged with developing an objective and reliable understanding of the perpetrators and processes leading to the event. The rational-actor perspective provides a framework through which to understand such acts – a framework that researchers can lean on to ensure that conclusions are not unknowingly biased by desperate horror. Such a view is not about relegating the suffering of victims. On the contrary, it recognises that it is sometimes too difficult to clearly see anything but the victims. Highlighting this perspective is merely acknowledging that outrage and horror can never be the sole foundation on which to build an accurate understanding of even the worst terrorist atrocities.

Focusing on the environment

Situational crime prevention has been one of the most important and successful strategies for combating different types of crime over the past 20 years. The situational approach is based on the understanding that the offence happens as a result of a number of decisions the offender has made, and that if actions are taken to affect these decisions then future offences can be prevented. The situational approach sees offenders as reasoning actors (i.e. the rational choice approach). The perspective simply holds that criminal activity – and this can certainly be expanded to terrorist activity – has an underlying rational basis to it.

The strongest support for the rational choice approach has come from studies that have looked at the activities and decision-making processes of professional property offenders. For example, Walsh (1980) interviewed 45 men in British prisons who had been convicted of burglary. Their ideal target was a business firm rather than a private house (more to be stolen) and, while half used information, the other half burgled on impulse (presumably, the latter were more likely to be caught and hence more likely to be available for interview). The greatest concern for the burglars was of being interrupted during the offence, as this increases the chances of violence occurring, and hence the likelihood of a more severe sentence if caught. The offenders saw themselves as desisting with increasing age (the risks of capture increased, sentences stiffened and the risk/return balance was generally less favourable).

Cornish and Clarke (1986) explicitly built into their models that criminals would never have access to all the information relevant to a particular incident. Ultimately, in any real-world situation, there are gaps in the knowledge of the offender. Clarke and Cornish also noted that there were clear limits to the resources offenders could devote to the planning, preparation and active decision-making for any one crime. Consider the following quote:

> Every offender must weigh, however quickly or imperfectly, whether the various economic and noneconomic benefits are worth the risks of arrest and confinement or retribution by the victim. Sometimes these decisions are made instantaneously, and at other times involve careful planning and searching.
>
> (Cornish and Clarke, 1986)

Criminals have limited resources available to them. There may be powerful pressures to complete the crime quickly; there may be poor access to the ideal equipment for the crime, or to the use of motor transport. Compounding these limits are the influence of factors such as the effects of alcohol or other drugs. Combined, these factors mean that decisions are often made under less than ideal conditions, with the result that it has always been relatively easy to find offences that appear to have been very poorly planned and executed. It also needs to be recognised that even for decision-makers possessing complete clarity of thought, the way in which different factors are assessed will almost always involve taking shortcuts in order to simplify the complex process of contrasting advantages with disadvantages (Sommers and Baskin, 1993).

Terrorist violence is typically well-planned and well-organised. As a result, several researchers have already noted the value of using a rational choice perspective in trying to understand and respond to terrorist planning and decision-making, and a number of research studies have already identified the rational basis behind terrorist decision-making and how these can identify possible points of intervention to prevent or disrupt effective operations (e.g. Taylor, 1988; Jacobs, 1998; Silke, 2003; Clarke and Newman, 2006).

Research such as that of Pape (2005) has highlighted that logic and reasoning also provide the foundation for suicide terrorism operations, and while potential suicide terrorists may not be deterred by the risk of being killed on a mission, they are certainly concerned about the risk of failure and the risk of being captured. Like other terrorists, suicide terrorists – and the support network and organisation around them – in planning for an attack are aiming for the most successful outcome possible at the least potential risk of failure. If the devices fail to detonate or the operatives are captured or killed before they can reach their targets or activate their devices, then the effort has failed.

Table 4.1 outlines the key factors that impact on how terrorist groups select targets for attacks.

Ideology and target selection

Ideology sets the wider context for terrorist groups. It establishes who the enemy is and what the organisation is fighting to achieve. Terrorist conflicts can be motivated by religious beliefs, nationalist and ethnic identities, and a wide range

Table 4.1 Key factors in terrorist target selection

1. Ideology
2. Internal factors to the terrorist group:
   - Material resources (weapons, money, etc).
   - Human resources (numbers, quality of leaders and members).
   - Decision-making process within the group.
3. External factors:
   - Target protection.
   - Security environment.
   - External opinion and events.
   - Nature of the society within which they operate.

of political doctrines. In order to gain insight into a group's ideology, a good start is to examine the writings and publications of the movement. For example, with regard to al-Qaeda, the writings of its chief spokesman and second-in-command, Ayman al Zawahiri, are highly informative:

> The first front is to inflict losses on the western crusader, especially to its economic infrastructure with strikes that would make it bleed for years. The strikes on New York, Washington, Madrid and London are the best examples for that.... In the second front, we have to get crusaders out of the lands of Islam especially from Iraq, Afghanistan and Palestine.... The third front is working at changing the corrupt regimes [in our countries], which have sold their honour to the Crusading west and befriending Israel.... The fourth front is popularizing the Dawah [inviting non-Muslims to accept the truth of Islam] work.
>
> (al Zawahiri, 2006)

This statement is very explicit in terms of the aims and objectives of the organisation, and also provides explicit insight into the types of targets the group is interested in (and why). Elsewhere in his writings, al Zawahiri talks explicitly about the type of tactics al-Qaeda and its affiliates should favour:

> [We] need to inflict the maximum casualties against the opponent, for this is the language understood by the west, no matter how much time and effort such operations take.

> [We] need to concentrate on the method of martyrdom operations as the most successful way of inflicting damage against the opponent and the least costly to the mujahideen in terms of casualties.

> The targets as well as the type and method of weapons used must be chosen to have an impact on the structure of the enemy and deter it.
>
> (al Zawahiri, 2001)

Such insight can be followed up by examining further documentation such as the instruction manuals many groups produce and use. One manual associated with al-Qaeda, *Military Studies in the Jihad*, states:

> The main mission for which the Military Organization is responsible is: The overthrow of the godless regimes and their replacement with an Islamic regime.
>
> Other missions consist of the following:
>
> 1 Gathering information about the enemy, the land, the installations, and the neighbours.
> 2 Kidnapping enemy personnel, documents, secrets, and arms.
> 3 Assassinating enemy personnel as well as foreign tourists.
> 4 Freeing the brothers who are captured by the enemy.
> 5 Spreading rumours and writing statements that instigate people against the enemy.
> 6 Blasting and destroying the places of amusement, immorality, and sin; not a vital target.
> 7 Blasting and destroying the embassies and attacking vital economic centres.
> 8 Blasting and destroying bridges leading into and out of the cities.

Having established an appropriate range of targets for violence, these manuals and other writings often follow this by providing instruction to members on how to plan, prepare for and then carry out attacks (e.g. giving instructions on how to construct weapons and carry out reconnaissance of potential targets).

One sees the same trends with other terrorist groups that come from very different contexts and backgrounds. For example, the IRA's instruction manual, *The Green Book*, outlines very clearly the overall strategy of the organisation for new members. This states that the IRA's aims are to carry out:

> A war of attrition against enemy personnel which is aimed at causing as many casualties and deaths as possible so as to create a demand from their people at home for their withdrawal.
>
> A bombing campaign aimed at making the enemy's financial interest in our country unprofitable while at the same time curbing long term financial investment in our country.
>
> To make the Six Counties as at present and for the past several years ungovernable except by colonial military rule.
>
> To sustain the war and gain support for its ends by National and International propaganda and publicity campaigns.

By defending the war of liberation by punishing criminals, collaborators and informers.

(Coogan, 1987: 693)

A terrorist motivated by left-wing ideology highlights how strategy – or 'politics' as he refers to it – is then translated into attacks:

The first decision is political – determining appropriate and possible targets. Once a set of targets is decided on, they must be reconnoitered and information gathered on how to approach the targets, how to place the bomb, how the security of the individuals and the explosives is to be protected. Then the time is chosen and a specific target. Next there was a preliminary run-through – in our case a number of practice sessions. Sometimes we don't do this as well as we should. The discipline during the actual operation is not to alter any of the agreed-upon plans or to discuss the action until everyone's safe within the group again. Our desire is not just for one success but to continue as long as possible.

(Worthy, 1970: 30)

The general process for preparing for and then carrying out attacks is quite similar across terrorist groups. Even though the last comments above are now 40 years old, they still apply as strongly to terrorists active today as they did in 1970. Naturally, there is variation in settings, resources and weapons between groups and this produces variation in targets. Some groups are more experienced and skilled than others and enjoy better resources. Not surprisingly their attacks tend to be more professional and sophisticated. Groups also tend to have preferred modus operandi – tactics and approaches with which they are experienced and use repeatedly even when other options are available (Silke, 2003).

In the material covered so far, nothing much has been said about targeting sporting events such as the Olympics. Both the IRA and al-Qaeda have targeted sporting events in the past as part of their campaigns of violence. At different times, such targets have been considered legitimate and hitting them was judged by the group as serving the movement's overall objectives. Having said that, it would be wrong to suggest that sporting targets are commonly or routinely selected for terrorist attack. Such attacks are not common, but when they do occur they can generate enormous media and public attention (as seen with the shooting of the Togo football team bus in Angola in January 2010 in advance of the start of the African Cup of Nations).

Ultimately terrorist groups are conscious that there are limits to what they can do. These limits are effectively imposed by their supporters and sympathisers. Certain tactics can be used and certain targets selected only provided that the support base for the movement supports such decisions. Consider, for example, this very telling comment from Eamon Collins, who for several years was an intelligence officer in the Provisional IRA:

> The IRA – regardless of their public utterances dismissing the condemnations of their behaviour from church and community leaders – tried to act in a way that would avoid severe censure from within the nationalist community; they knew they were operating within a sophisticated set of informal restrictions on their behaviour, no less powerful for being largely unspoken.
>
> (Collins, 1997: 296)

The implication is that the terrorists must behave in an acceptable manner or else they will be rejected by their supporters (and potential supporters) and thus ruined. Comments such as Collins' above clearly indicate that this is a fact very much to the forefront of terrorist thinking. Even al-Qaeda, an organisation which has shown itself to be repeatedly willing to kill very large numbers of civilians in attacks, operates within such a set of restrictions. Consider the following extracts from another letter written by Ayman al Zawahiri to another terrorist commander criticising the excessive violence of some of that commander's attacks and actions:

> the mujahed movement must avoid any action that the masses do not understand or approve ... the general opinion of our supporter does not comprehend that.... And we should spare the people from the effect of questions about the usefulness of our actions in the hearts and minds of the general opinion that is essentially sympathetic to us.... I say to you: that we are in a battle, and that more than half of this battle is taking place in the battlefield of the media. And that we are in a media battle in a race for the hearts and minds of our Umma ... [we must not expose] ourselves to the questions and answering to doubts [among the Umma]. We don't need this.

Zawahiri himself has ordered some extremely ruthless acts in his time, but he was trying to drive home the damaging impact of the other commander's attacks in terms of undermining wider support from the movement.

While targeting sporting events can be a sensitive issue – as the IRA found when a hoax bomb they planted resulted in the 1997 Grand National being cancelled – there is no sense that this means such attacks are specifically 'off-limits' (McGladdery, 2006). The IRA might hesitate before targeting the Grand National again, but al-Qaeda – with an established track record of mass casualty attacks carried out at 'soft' civilian locations – does not have such inhibitions. Indeed, there is good reason to believe that al-Qaeda clearly sees enormous potential benefits in hitting major sporting events and especially ones as large as the Olympics. For terrorists, an attack that receives a great deal of media attention is usually seen as much more successful than an attack that receives relatively little (even if the human casualties and physical damage caused by both attacks are similar). Indeed, even if an attack results in the death or capture of all the terrorists involved, it can still be regarded as highly successful if it has received intense international media attention. Consider the following assessment from Abu 'Ubeid Al Qurashi, an al-Qaeda activist who was assessing the impact of the Palestinian attack on the 1972 Munich Olympics:

Seemingly, the [Munich Olympics] operation failed because it did not bring about the release of the prisoners, and even cast a shadow of doubt on the justness of the Palestinian cause in world public opinion. But following the operation, and contrary to how it appeared [at first], it was the greatest media victory, and the first true proclamation to the entire world of the birth of the Palestinian resistance movement.... In truth, the Munich operation was a great propaganda strike. Four thousand journalists and radio personnel, and two thousand commentators and television technicians were there to cover the Olympic Games; suddenly, they were broadcasting the suffering of the Palestinian people. Thus, 900 million people in 100 countries were witness to the operation by means of television screens. This meant that at least a quarter of the world knew what was going on in Munich; after this, they could no longer ignore the Palestinian tragedy.... There are data attesting to the importance of the Munich operation in the history of the resistance movement, and the extent of its influence on the entire world. It is known that a direct consequence of this operation was that thousands of young Palestinians were roused to join the *fedayeen* organizations.... The number of organizations engaging in international 'terror' increased from a mere 11 in 1968 to 55 in 1978. Fifty-four percent of these new organizations sought to imitate the success of the Palestinian organization – particularly the publicity the Palestinian cause garnered after Munich.

(Al-Qurashi, 2002)

Thus, from an al-Qaeda perspective, there are potentially many benefits to targeting the Olympics. In the UK, convictions have already occurred of individuals on terrorism charges where the individuals talked about targeting the 2012 Games. These individuals were arrested and convicted long before such talk could be turned into action, but it would be naive to think they were the only ones to hold such views, and that al-Qaeda no longer viewed the Games as an appropriate target. Indeed, in his own assessment, Al-Qurashi judged that only one other terrorist attack in the past 40 years could be judged to have been more successful than the 1972 Munich attack: the 9/11 attacks.

The rationales for others who targeted the Olympics are often similar, even if the ideological motive behind the attack is radically different. Consider, for example, the reasons given by Eric Rudolph who was responsible for the bomb attack against the Atlanta Olympics in 1996:

Abortion is murder. And when the regime in Washington legalized, sanctioned and legitimized this practice, they forfeited their legitimacy and moral authority to govern.... In the summer of 1996, the world converged upon Atlanta for the Olympic Games. Under the protection and auspices of the regime in Washington millions of people came to celebrate the ideals of global socialism. Multinational corporations spent billions of dollars, and Washington organized an army of security to protect these best of all games. Even though the conception and purpose of the so-called Olympic movement is to promote the values of global socialism, as perfectly expressed in the song 'Imagine' by John Lennon, which was the theme of the 1996 Games even though the purpose of the Olympics is to promote these despicable ideals, the purpose of the attack on July 27 was to confound, anger and embarrass the Washington government in the eyes of the world for its abominable sanctioning of abortion on demand.

The plan was to force the cancellation of the Games, or at least create a state of insecurity to empty the streets around the venues and thereby eat into the vast amounts of money invested.

(Rudolph, 2005)

Examining the rationales for these terrorist attacks reveals a further critical issue. By and large, terrorists have targeted the Olympics not because they bear any particular ill-will to sport in general or the Olympics in particular, but rather because they wish to target a government involved in the Games. In the vast majority of cases, it is the host government who is the real and primary target for any attack. Because of its extremely high profile, the Olympics is being used as a way to embarrass and humiliate the host government in a highly public fashion. The host government may also suffer potentially serious economic, political and psychological impacts as a result of a successful attack. Less common are attacks that are specifically targeting one of the nations competing in the Olympics. The 1972 Munich attack is a clear example of this scenario. Here, the national team is used by the terrorists as a proxy to represent their whole nation.

The intense national and international media interest also helps to explain another finding with regard to the Olympics: the vast majority of terrorist attacks and plots target the Summer Games, with very few targeting the Winter Olympics. A key factor in this is that there is simply more media and public interest in the Summer Games than the Winter Games. More people watch media coverage of the former and watch it for longer than is the case for the latter. Table 4.2 gives a breakdown of the viewing figures for some recent Olympics.

Thus viewing figures for the Summer Games are roughly three times higher than those for the Winter Games. Other figures also swiftly demonstrate the appeal of the Olympics over other major sporting events in this regard. Table 4.3 shows the most-watched sporting events from 2008.

As we can see, the top viewing figures for the Olympics were between four- and five-times larger than the viewing figures for even the biggest other sporting

Table 4.2 Media viewing figures for the Olympics

Summer Olympics		Winter Olympics	
Total viewer hours		Total viewer hours	
Sydney 2000	36.1 billion	Salt Lake City 2002	13.1 billion
Athens 2004	34.4 billion	Torino 2006	10.6 billion

Source: International Olympic Committee (2010).

Table 4.3 Most-watched sporting events in 2008

Event	Viewers
Beijing Olympics Opening Ceremony	984 million*
Beijing Olympics Closing Ceremony	778 million
Euro 2008 Final (Germany vs Spain)	287 million
Champions League Final (Manchester United vs Chelsea)	208 million
Super Bowl (New York Giants vs New England Patriots)	152 million

Source: Harris, 2009.

Note
*Some estimates place this figure closer to 1.2 billion.

events. This huge discrepancy in media and public interest lies behind the particular interest in targeting the Games compared to the efforts made to attack other sporting events. This is not to suggest that other events are immune from attack. They are not, and attacks do occur. However, they do not occur with anything like the frequency and scale of the plots directed against the Olympics.

Linked to this are concerns with regard to the Paralympics. While researching for this book, this is a question I have been asked by police officers and others involved in preparations for the 2012 Games: how likely is it that terrorists would attempt to attack the Paralympics? The short answer is that it is not particularly likely that the Paralympics will be directly targeted by terrorists and there are two reasons for this. First, wider media interest in the Paralympics is vastly lower than the media interest seen for the Olympics. At a fundamental level, the potential reward in terms of media exposure and impact is greatly reduced and so the incentive for an attack is also greatly reduced. Second, and arguably more important, terrorist groups are unlikely to view attacking disabled athletes as a saleable decision in terms of their supporters and constituents. As al Zawahiri warned earlier, the terrorist group's wider support base needs to be able to understand and agree with any acts of violence: the 'movement must avoid any action that the masses do not understand or approve'. Fundamentally, the terrorists are aiming to hurt a government, and it would be difficult to build a convincing argument that an attack on the Paralympics does this in any meaningful way. In contrast, it is much easier to build a case that an attack on the Olympics does. Consequently the Paralympics themselves are unlikely to be deliberately targeted.

While terrorist ideologies can differ hugely, the general manner in which the attacks themselves are planned, prepared and then launched is more or less the same across groups. The instructions given in terrorist manuals on how this should be done are essentially the same. Drake (1998) has identified several stages that are typical of most terrorist operations. These are outlined in Table 4.4.

Table 4.4 Stages of a terrorist operation

1	Setting up logistical network (safe houses, arms dumps, vehicles, ID, etc.). Partially dependent upon how overt or covert the terrorists' ordinary lives have to be.
2	Selection of potential targets – generally based on ideological and strategic considerations.
3	Information gathering on potential targets.
4	Reconnaissance of potential targets (find out the likely location and protection of the target at a specific time).
5	Planning of the operation.
6	Insertion of weapons into the area of operation.
7	Insertion of operators into the area of operation.
8	Execution of the operation.
9	Withdrawal of the operational team (not applicable if successful suicide attack).
10	Issue of communiqués if appropriate.

Note
Adapted from Drake (1998).

As Table 4.4 highlights, the preparatory phase of an operation is by far the longest stage with the most elements. Drake noted that some of these earlier stages, such as setting up the logistical network, or inserting weapons into an area, may occur long before a target is chosen, but may influence the choice of targets. Also, the degree of planning that takes place for an attack can vary. Thus, a pre-planned operation may involve meticulous preparation, the careful assembly of resources and personnel, and a well-defined plan of action. On the other hand, an opportunist operation may simply occur because a terrorist is suddenly confronted with an opportunity to attack a suitable target and has the means available to do so. Such a spontaneous scenario is less likely in the context of attacks against major high-profile targets such as the Olympics, although it is worth highlighting that the two most successful terrorist attacks against the Olympics – Munich and Atlanta – both involved relatively modest levels of planning. Serious planning for the Munich attack only started in mid-July 1972, less than seven weeks before the actual assault was launched. The decision to try to attack the Olympics was first proposed at a meeting in Rome on 15 July, and two days later a terrorist was sent to Munich to conduct the first reconnaissance of the Olympic site (Wolff, 2002). The Atlanta attack was pulled together at even shorter notice – just a few weeks before the attack was launched. The implication for security planners is that many plots will be pulled together at very short notice, allowing a very small window of opportunity to detect the preparations for any assault.

The tactics, planning, preparation and target selection of the terrorists is limited by the capability of the terrorist group. Drake (1998) highlighted the fact that key aspects of the group's capability can be broken down into five elements:

Quality of leadership – leaders are crucial for effective organisation, recruitment, group motivation, refining and improving tactics and strategies. Leaders also play a critical role in highlighting appropriate targets and tactics for the movement as a whole.

Quantity of members – the larger the number of members the group has, the more ambitious the group can be in terms of the number of operations it can mount (perhaps simultaneously), the complexity of these operations, and the level of intelligence and information the group can draw upon in planning and preparation. Greater numbers, though, also increase the risk of detection and infiltration by the security forces.

Quality of members – the more experienced, better-skilled, better-trained and more highly motivated members are, the greater the capability the group possesses.

Weapons available – determines the types of target and tactics that are feasible.

Financial base – the greater the funds available, the more sophisticated the types of operation a terrorist group can undertake. A lack of funds can force the group to divert effort into fund-raising activities (with accompanying risks and distractions).

As a result of real limits in capabilities – and with finite resources in time, money, information, skills and expertise – terrorist planning and preparation will never be perfect. Indeed, even dramatically successful terrorist attacks have still displayed failures and misjudgements by the perpetrators in the planning and execution stages (e.g. the failure of Flight 93 to reach its target on 9/11).

The impact of such limits is well-illustrated by the bombing that targeted the Atlanta Olympics in 1996. Following his eventual capture and conviction for the bombing, Eric Rudolph provided a statement giving some insight into his planning and decision-making with regard to the attack:

> The plan was conceived in haste and carried out with limited resources, planning and preparation.... Because I could not acquire the necessary high explosives; I had to dismiss the unrealistic notion of knocking down the power grid surrounding Atlanta and consequently pulling the plug on the Olympics for their duration.
>
> The plan that I finally settled upon was to use five low tech timed explosives to be placed one at a time on successive days throughout the Olympic schedule.... The attacks were to have commenced with the start of the Olympics, but due to a lack of planning this was postponed a week.
>
> (Rudolph, 2005)

In the end, only one device was planted, although this resulted in two deaths and over 100 injured. After this attack, Rudolph decided against carrying out further attacks as planned and destroyed the other four devices.

Appreciating that offenders, when making decisions, attempt to minimise the likelihood of failure and maximise the likelihood of success allowed criminologists to develop a range of techniques to deter and prevent crimes from occurring. Cornish and Clarke (2003) ultimately produced a list of 25 separate techniques for situational crime prevention. These techniques were grouped in five themes:

- Increase the effort the offender has to make.
- Increase the risks the offender faces.
- Reduce the rewards the offender will get if successful.
- Reduce provocations that might encourage offending.
- Remove excuses that might justify offending.

The specific techniques are listed in Table 4.5.

Not all of these techniques are equally applicable to preventing and deterring terrorism, though many are already playing a major role in government strategy for meeting such a threat (e.g. target hardening). Clarke and Newman (2006)

Table 4.5 Techniques of situational crime prevention

Theme	Technique
Increase the effort	Target harden Control access to facilities Screen exits Deflect offenders Control tools/weapons
Increase the risks	Extend guardianship Assist natural surveillance Reduce anonymity Utilise place managers Strengthen formal surveillance
Reduce the rewards	Conceal targets Remove targets Identify property Disrupt markets Deny benefits
Reduce provocations	Reduce frustrations and stress Avoid disputes Reduce emotional arousal Neutralise peer pressure Discourage imitation
Remove excuses	Set rules Post instructions Alert conscience Assist compliance Control drugs and alcohol

recently considered how the model could be used to target suicide terrorism in particular. To do this, they examined data on suicide attacks carried out in Israel from 2000 to 2002. They then produced a substantial list of potential interventions drawing on the situational perspective. One of the issues that quickly became apparent is that not all of the recommendations would be suitable or relevant outside the Israeli context. For example, the issues raised with regard to payments made by terrorist organisations to the families of suicide bombers are not relevant to al-Qaeda or its affiliates as the organisation does not attempt to support families in this way.

Other recommendations would probably be too expensive to implement in most environments. For example, in Israel during the period considered in the Clarke and Newman study, there were 60 successful suicide attacks in the space of a year, averaging at roughly one successful attack every ten days. This was a very high-intensity campaign and one that would justify many of the measures advocated, including very significant disruptions to public and economic life (e.g. altering public-transport practices at short notice, closing roads, security checks at restaurant entrances, etc.).

In contrast, it would be much more difficult to endorse similarly extensive measures in a UK context, where there has been only one successful suicide attack in the past ten years. This issue was well-illustrated by the significantly increased airport security regulations introduced (temporarily) in August 2006 in response to fears of an imminent terrorist attack. There are significant economic costs to such measures, and in the absence of clear evidence of regular attempted attacks, long-term support for such measures can be expected to be very fragile indeed. In contrast, for environments where suicide attacks are relatively common (e.g. Israel, Iraq, Sri Lanka, etc.) support for more wide-ranging measures is likely to be robust.

The Olympics presents something of a different scenario because of its extremely high profile and the clear history of disparate terrorist groups attempting to carry out attacks to disrupt or coincide with the Games. Within the context of 2012, for example, security planners know that the Games and the period immediately before and after are especially high-risk for attacks (the most high-risk period is the very start of the Games which is the time most terrorist plots aim to strike). As a result, this sanctions the spending of a great deal of money on security-related efforts that ordinarily would not be the case. At the time of writing, the security budget for 2012 is estimated at £710 million, but could well rise closer to £1.5 billion (and potentially could reach even higher). A key question at this stage is: how effective will these measures be in preventing or deterring attacks?

Target hardening and terrorism

Prevention is widely regarded as the most cost-effective approach in dealing with terrorism, and certainly underlies the situational approach advocated by Clarke, Newman and others. Once a terrorist campaign has established itself, however, responding becomes very costly. An impact evaluation of Northern Irish terrorism, for example, found that for every £1 the paramilitaries were able to raise in finances, it ultimately cost the governments attempting to combat them *at least* £130 (Silke, 1998). Similarly, a recent review of the conflict in Iraq estimates the cost to the US alone as being at around $3 trillion and rising. Such figures help to illustrate the critical threat posed by terrorism to a state.

Given such high costs, it is not surprising that a lot of effort is invested in deterring and preventing terrorist attacks. Target hardening has become one of the key tools in these efforts. Target hardening can take a variety of forms, ranging from what is termed 'multi-event thwarting' to 'single-event thwarting'. Single-event thwarting refers to something such as installing metal detectors in airports. This is a measure that is meant largely to prevent just one type of terrorist activity (i.e. the hijacking of aircraft) at one specific location. Multi-event measures, however, are more general and include something like the creation of a highly-trained commando force to use in terrorist scenarios. The commandos can be used in a wide variety of situations and can carry out a range of different functions: dealing with siege situations, to free hostages, to protect VIPs, to ambush terrorists, etc.

One problem with target hardening in counter-terrorism is the very real risk of displacement. In short, when you make it more difficult for terrorists to carry out one type of attack, you increase the chance that they will start trying to carry out new types of attack instead. The displacement effect was first seen by criminologists who were studying crime in cities. The criminologists found when target hardening measures were introduced in one area (e.g. CCTV cameras, extra security guards, better designed barriers and locks, etc.), crime levels did indeed drop. However, the bad news was that crime rates in neighbouring areas which had not received the new target hardening increased! For certain types of crime, the displacement was almost total: crime dropped in some areas but the crime rate for the city as a whole more or less stayed the same.

This displacement effect has also been seen with efforts to combat terrorism. One of the clearest examples comes from the introduction of metal detectors in airports in the 1970s. Statistics on terrorist hijackings show that there was a massive (and sustained) decrease in skyjackings once the metal detectors went in. However, this was accompanied by a massive increase in other types of attacks, such as assassinations and kidnappings (Cauley and Im, 1988; Enders *et al.*, 1990). While technology has since been developed to circumvent metal detectors, terrorists have been slow to exploit this. Instead, in the relatively few cases where hijackings have succeeded in recent decades, it is because the airports do not have metal detectors (or those that did exist were not being used properly), or else because the terrorists have been able to exploit lax security regulations that have allowed them to bring potential weapons on board (e.g. the 9/11 hijackers).

In a similar vein, increased fortification of US embassies in 1976 reduced attacks against these buildings but resulted in a substitution into assassinations. Instead of attacking the embassies, the terrorists started going after embassy staff in their homes or while they were travelling to work (Enders and Sandler, 1993). Even worse, sometimes significantly beefing up security seems to have not had any impact whatsoever. Research showed that increased funding for embassy security in 1986 had no deterrent effect on terrorism – despite the fact that over $2.4 billion was spent on security improvements. The researchers drew attention to some of the critical implications emerging from these findings:

> A number of policy insights can be drawn. First, the unintended consequences of an antiterrorism policy may be far more costly than the intended consequences, and must be anticipated. In the case of metal detectors, kidnappings increased; in the case of embassy fortification, assassinations became more frequent. Protected persons may have faced more life-threatening attacks owing to security measures. By cutting down on threats, metal detectors may have induced terrorists to substitute deeds for words. Second, piecemeal policy, in which a single attack mode is considered when designing antiterrorism action, is inadequate. This follows because the various attack modes are interrelated through substitutability and complementarity.
>
> (Enders and Sandler, 1993: 843)

Ultimately, some situational prevention strategies will have a positive impact on attacks against particular targets. Research has shown that terrorist attacks against such targets often show a significant and long-term decrease in the aftermath of improved security. However, this is frequently accompanied by an increase in attacks against other targets (e.g. interventions to improve US embassy security in the 1980s were followed by a long-term decrease in attacks against the embassies, but were accompanied by a long-term increase in attacks against embassy staff outside of the embassy building). Such displacement effects have often been found to be very high.

Within the context of the UK, a good example of the benefits and potential risks of target hardening comes from the so-called 'Ring of Steel' in London (Silke, 2008). The 'Ring' was composed of heightened and extensive security improvements around London's major financial district and was introduced after the IRA started to attack the area in 1992. The first attack took place on 10 April 1992, when an IRA team parked a van filled with explosives close to the Baltic Exchange in the City of London (equivalent to the stock exchange on Wall Street in New York). The IRA issued a warning about the device and the surrounding area was largely evacuated before the massive bomb exploded. Even so, three people were still killed and the structural damage to surrounding buildings was absolutely massive. For the IRA, the attack was seen as the most successful and important operation they had carried out that year. The economic impact of the attack was staggering – estimated at over £800 million. To help further highlight the significance of the bombing, it is worth noting that the combined economic impact of 10,000 IRA bomb attacks in Northern Ireland over the preceding 23 years was just £615 million (McGladdery, 2006). Incredibly, the Baltic

Exchange attack had caused the UK more economic harm than *all* previous IRA bombings combined. It was a powerful incentive for the IRA to continue with the new approach.

In the aftermath, a so-called 'Ring of Steel' was created around the financial district in the City of London. The aim of this was to protect this vital district from further attacks and to deter any additional bombings. Millions were spent installing a sophisticated CCTV system, introducing new barriers, altering transport flow into and through the area, introducing random vehicle searches, increasing police checkpoints as well as a host of other new measures.

The transformed security arrangements in the area received wide publicity and this is an important factor. In order to be deterred, people have to be aware of increased security. Thus, most major security improvements involve a media element so that the public – and hence the terrorists – are aware that new measures are in place. The extensive target hardening certainly made it much more difficult for the IRA to carry out an attack in the area, but not impossible.

The IRA studied the new security arrangements and found a flaw. They discovered that large vehicles linked to construction work in the protected area were allowed into the protected district and were able to park in restricted places on Saturday mornings. On 24 April 1993, the IRA used a truck to carry a 2,300 lb fertiliser bomb into Bishopsgate, the financial heart of London. The resulting blast was much bigger than the Baltic Exchange bombing the previous year and the explosion caused damage approaching $2 billion to buildings, housing and prestige foreign and domestic groups like Nat West Bank, Hong Kong and Shanghai Bank, Barclay's Bank and the Abu Dhabi Investment Bank. The impact was so severe that it brought about a crisis in the insurance industry and led to the near collapse of Lloyds of London. One person was killed in the explosion and over 60 were injured. Yet again, the operation was a huge success for the IRA, showing them as an organised and highly effective group with a real capacity to critically hurt the UK's economy, and was almost certainly the highpoint of their campaign that year.

For London, it was a hard lesson that target hardening does not always work. When new measures are introduced, terrorists will study these and attempt to exploit loopholes. Nevertheless, after Bishopsgate, the 'Ring of Steel' was enhanced even further. New safety measures were added and, over time, the protected area was gradually expanded to cover even more of London. The new security regime had learned from the lessons of the past and was even tougher to penetrate than before. The IRA's response was to hit just outside the protected area.

On 9 February 1996, the IRA exploded a large truck bomb in Canary Wharf in London. This was another important financial area and indicated the IRA's desire to go after targets that would have a particularly heavy impact on the economy. Two people were killed and over $200 million worth of damage was caused. Four months later, the IRA detonated yet another large truck bomb but this time outside London entirely, striking at Manchester's city centre. The massive explosion caused $800 million worth of damage.

The lesson from the 'Ring of Steel' response was, first, you can (eventually) deter specific types of attacks from a particular location. However, terrorists will look for vulnerabilities in new defences and, if they find them, they'll exploit them. Second, other locations that have not benefited from improved security become increasingly attractive to terrorist planners. This was particularly obvious with the attack against Canary Wharf – a location which offered many of the same benefits in terms of impact as the Baltic Exchange and Bishopsgate but which lacked the formidable defences provided by the 'Ring of Steel'.

The 'Ring of Steel' was primarily designed to stop highly destructive vehicle bombs, but we also have examples that are explicitly focused on the threat of suicide terrorism. In recent years, one of the most telling examples of the impact of target hardening comes from Israel, where the construction of a major barrier has been a key element in the fight against suicide terrorism. Construction on the 450 mile long barrier began in 2002 at the height of the second Intifada when Palestinian suicide attacks were occurring on a weekly basis. From the beginning, the barrier has been deeply controversial and has been heavily criticised on legal, diplomatic and humanitarian grounds. Nevertheless, the expansion of the barrier has been clearly linked with a marked decline in successful suicide attacks. As Figure 4.1 shows, there were 60 suicide attacks in 2002 when construction started. This figure dropped steadily in each subsequent year, and in 2008 there was only one successful suicide attack.

It is difficult to argue against the positive impact the barrier has made in preventing suicide attacks. The effort, however, has not been without its downside. As already indicated, the barrier has been very controversial for a range of reasons. As a construction project, it has also been plagued by problems and is desperately behind schedule (the original finish date was in 2004 but work is still

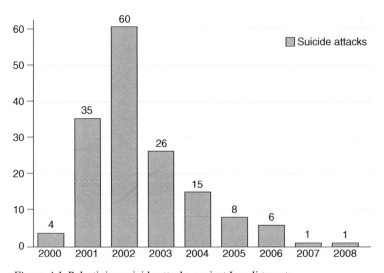

Figure 4.1 Palestinian suicide attacks against Israeli targets.

not complete and it is very likely that the barrier will never be finished along the original plans). Long delays have added to the costs, which have proven to be far greater than originally anticipated. The project has so far cost at least $1.5 billion, and construction work has halted at different times when funding has run out.

The barrier is also a lesson in the nature of displacement. While it has certainly been a success in dramatically reducing suicide attacks, this has not meant that Palestinian militants have given up trying to carry out attacks against Israel. Instead, we have seen a steady shift into other types of attack that are not hindered by the barrier. Figure 4.2 shows the huge growth in rocket attacks, which have increased massively just as suicide attacks have been declining.

The good news from the Israeli perspective is that a rocket attack is far less accurate than a suicide bomber, and even though there have been literally thousands of rockets fired into Israel in recent years, they have so far caused only a handful of deaths.

Conclusions

Target hardening interventions can have a real and positive impact on preventing terrorist attacks. Research has shown that terrorist attacks against such targets generally show a significant and long-term decrease in the aftermath of improved security. However, this is accompanied by an increase in attacks against other targets (e.g. interventions to improve US embassy security in the 1980s were followed by a long-term decrease in attacks against the embassies, but also a long-term increase in attacks against embassy staff outside the embassy building). Such displacement effects have been found to be almost 100 per cent – the

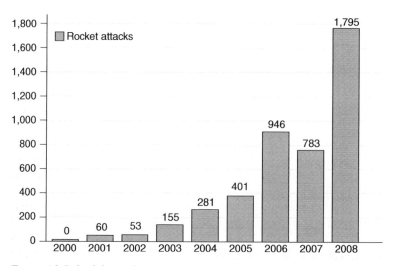

Figure 4.2 Palestinian rocket attacks against Israeli targets.

terrorists do not slacken off in their commitment to carry out attacks (Enders and Sandler, 2003).

As the lessons from London's 'Ring of Steel' and Israel's defence barrier have shown, terrorists will try to find and exploit vulnerabilities in new security measures. If they cannot do this, they will then switch their attacks to a new location or else will adopt entirely new tactics that are not thwarted by the security measures.

Such displacement effects have already been seen with regard to terrorist attacks targeting the Olympics. Very few terrorist attacks have actually occurred within the sporting venues themselves. Indeed, since 1972, security around such locations has usually been very good and plots aimed at penetrating this security have largely failed. This, however, has not convinced terrorist groups to abandon efforts to carry out acts of violence. The intense media interest in the Olympics has convinced terrorists that simply striking close to Olympic venues will still garner enormous publicity and impact. Thus, most plots have not focused on penetrating the increasingly formidable security screens that protect the main sites.

For planners, this poses a significant challenge. Clearly the priority has to be to protect the main sites and venues. However, experience shows that such protection does not deter terrorists from carrying out attacks in connection with the Games. Looking to 2012, for example, would anyone consider it acceptable that no attack took place at an Olympic venue but that terrorist bombs were detonated at a shopping centre elsewhere in London? The terrorists could expect to gain almost as much media attention for such an attack as if they had detonated a bomb at an Olympic venue. The difficulties they would face in carrying out the shopping centre attack would be considerably lower.

To help deal with the problem of displacement, researchers have suggested that emphasis should be given to multi-event thwarting initiatives (Sandler *et al.*, 1987). Single-event thwarting refers to something such as installing metal detectors in airports. This is a measure that is meant to prevent just one type of terrorist activity (i.e. the hijacking of aircraft). Multi-event measures include something like the creation of a commando force to use in terrorist scenarios. For example, in the aftermath of the Munich debacle, the West German government authorised the creation of an elite commando unit, GSG-9, which could develop the skills, equipment and tactics to resolve future threats posed by terrorists (Koch and Hermann, 1977). Such commando units can be used in siege situations, to free hostages, to protect VIPs, to ambush terrorists, etc. The research concluded that, because of their adaptability, multi-event measures would be 'more effective in countering the substitution phenomenon' (Sandler *et al.*, 1987: 386).

Thinking along these lines is also needed in terms of Olympic security planning. Displacement effects are always likely in the context of very high profile and well protected targets, and this is perhaps the most critical challenge facing security planners at the Olympics and other major events. Considering what will happen and what should be done if the terrorists decide to strike a less protected target in the same city or region is the next step. The terrorists will always search for potential weaknesses in existing defences and plans, and those responsible for security must do the same.

References

Al-Qurashi, Abu 'Ubeid, (2002). 'Al-Ansar: For the Struggle Against the Crusader War,' Issue 4, February 27, trans. MEMRI, 'Al-Qa'ida Activist, Abu 'Ubeid Al-Qurashi: Comparing Munich (Olympics) Attack 1972 to September 11,' *Special Dispatch Series*, No. 353, March 12. Online, available at: http://memri.org/bin/articles.cgi?Page=archives&Area=sd&ID=SP35302 (accessed 14 February 2008).

Al Zawahiri, A. (2001). 'Knights Under the Prophet's Banner.' In Mansfield, L. (ed., 2006). *His Own Words: a Translation of the Writings of Dr Ayman al Zawahiri*. Old Tappan: TLG Publications.

Al Zawahiri, A. (2005). Letter to al-Zarqawi on 11 October. In Mansfield, L. (ed., 2006). *His Own Words: a Translation of the Writings of Dr Ayman al Zawahiri*. Old Tappan: TLG Publications.

Al Zawahiri, A. (2006). Statement on 4 March 2006. In Mansfield, L. (2006). *His Own Words: a Translation of the Writings of Dr Ayman al Zawahiri*. Old Tappan: TLG Publications.

Cauley, J. and Im, E.I. (1988). 'Intervention Policy Analysis of Skyjackings and Other Terrorist Incidents.' *American Economic Review*, 78, 2, 27–31.

Clarke, R. and Cornish, D. (1985). 'Modelling Offender's Decisions: a Framework for Research and Policy.' In Tonry, M. and Morris, N. (eds), *Crime and Justice: An Annual Review of Research*, Vol. 7. Chicago: University of Chicago Press.

Clarke, R. and Newman, G. (2006). *Outsmarting the Terrorists*. London: Praeger Security International.

Collins, E. (with McGovern, M.) (1997). *Killing Rage*. London: Granta Books.

Coogan, T.P. (1987). *The IRA*. London: Fontana.

Cornish, D. and Clarke, R. (1986). *The Reasoning Criminal*. New York: Springer-Verlag.

Cornish, D.B. and Clarke, R.V. (2003). 'Opportunities, Precipitators and Criminal Decisions: a Reply to Wortley's Critique of Situational Crime Prevention.' In Smith, M.J. and Cornish, D.B. (eds), *Theory for Practice in Situational Crime Prevention: Crime Prevention Studies*, Vol. 16. Monsey: Criminal Justice Press, pp. 41–96.

Drake, C. (1998). *Terrorists' Target Selection*. London: Macmillan.

Enders, W. and Sandler, T. (1993). 'The Effectiveness of Antiterrorism Policies: a Vector–Autoregression–Intervention Analysis.' *American Political Science Review*, 87, 4, 829–844.

Enders, W. and Sandler, T. (2003). 'What Do We Know About the Substitution Effect in Transnational Terrorism?' In Silke, A. (ed.), *Research on Terrorism: Trends, Achievements and Failures*. London: Frank Cass, pp. 57–71.

Enders, W., Sandler, T. and Cauley, J. (1990). 'UN Conventions, Technology and Retaliation in the Fight Against Terrorism: an Econometric Evaluation.' *Terrorism and Political Violence*, 2, 1, 83–105.

Gilmour, R. (1998). *Dead Ground: Infiltrating the IRA*. London: Little, Brown & Company.

Harris, N. (2009). '1,000,000,000: Beijing Sets World TV Record.' *The Sunday Times*, 10 May. Online, available at: www.timesonline.co.uk/tol/sport/olympics/article6256063.ece.

International Olympic Committee (2010). *Olympic Marketing Fact File* (2010 edition). Online, available at: www.olympic.org/Documents/IOC_Marketing/IOC_Marketing_ Fact_File_2010%20r.pdf.
Jacobs, S. (1998). 'The Nuclear Threat as a Terrorist Option.' *Terrorism and Political Violence*, 10, 149–163.
Koch, P. and Hermann, K. (1977). *Assault at Mogadishu*. London: Corgi.
McGladdery, G. (2006). *The Provisional IRA in England: the Bombing Campaign 1973–1997*. Dublin: Irish Academic Press.
Military Studies in the Jihad Against the Tyrants (n.d.). Online, available at: http://cryptome.org/alq-terr-man.htm#6 (accessed 13 April 2010).
Pape, R. (2005). *Dying to Win: the Strategic Logic of Suicide Terrorism*. New York: Random House.
Rudolph, E. (2005). 'Statement at Sentencing Concerning the Centennial Park Bombing.' Online, available at: www.ericrudolph.org/centennialpark.htm (accessed 15 April 2010).
Sandler, T., Atkinson, S., Cauley, J., Im, E., Scott, J. and Tschirhart, J. (1987). 'Economic Methods and the Study of Terrorism.' In Wilkinson, P. and Stewart, A.M. (eds), *Contemporary Research on Terrorism*. Aberdeen: Aberdeen University Press, pp. 376–389.
Silke, A. (1998). 'In Defence of the Realm: Financing Loyalist Terrorism in Northern Ireland – Part One: Extortion and Blackmail.' *Studies in Conflict and Terrorism*, 21, 4, 331–362.
Silke, A. (2001). 'When Sums Go Bad: Mathematical Models and Hostage Situations.' *Terrorism and Political Violence*, 13, 2, 49–66.
Silke, A. (2003). 'Beyond Horror: Terrorist Atrocity and the Search for Understanding – The Case of the Shankill Bombing.' *Studies in Conflict and Terrorism*, 26, 1, 37–60.
Silke, A. (2008). 'Target-Hardening and Terrorism: Challenges and Risks.' *Journal of Counterterrorism and Homeland Security International*, 14, 2, 26–28.
Sommers, I. and Baskin, D. (1993). 'The Situational Context of Violent Female Offending.' *Journal of Research in Crime and Delinquency*, 30, 2, 136–162.
Taylor, M. (1988). *The Terrorist*. London: Brassey's.
Walsh, D. (1980). *Break-Ins: Burglary from Private Houses*. London: Constable.
Wolff, A. (2002). 'When the Terror Began.' *Time*, 25 August 25. Online, available at: www.time.com/time/europe/magazine/2002/0902/munich/index.html.
Worthy, W. (1970). 'A Real Bomber's Chilling Reasons.' *Life*, 68/11, 27 March, p. 30.

Part II
Response themes

Transport security, the role of surveillance and designing stadia for safer events

5 Securing the transport system

Steve Swain

Introduction

An effective and efficient transportation system is vital for any city hosting the Olympics in order to get people into the country, move them around the host city and transport them to and from the Olympic venues. Thus any attack that shuts down the transport network has a direct impact on the Games and arguably is an attack on the Olympics itself, regardless of whether the actual venue is targeted or not. This chapter outlines the four primary types of transportation system, road, rail, air and marine, that would be used to move people to and from an Olympic venue, both from outside the country and internally. The discussion is developed by examining the various methodologies that may be deployed to protect such a system from terrorist attack and their relative effectiveness and ability to protect the travellers on the different systems.

A total of 2,676 people were killed in transport accidents in the UK in 2008 (averaging at approximately eight deaths per day) (Office for National Statistics, 2009) and, although these deaths are tragedies for the families involved, they barely feature in the public consciousness. The Pan American 747 flight that was destroyed over Lockerbie in December 1988 as a result of a terrorist bomb killed 270 people, and the repercussions are still making headlines over 20 years later. Similarly, the London bombings of July 2005, where 52 people were killed, also feature in the media with persistent regularity. To give some context to these figures, the police investigation for 7/7 took 14 days to gather the evidence from the scene before one of the damaged trains was removed from the tunnel for further forensic examination: 14 days in which around 112 people were killed on UK roads. The author spent ten days working in Scotland on the Lockerbie investigation and during this period there were approximately 80 people killed on UK roads. While the numbers of these road deaths hardly seem to feature in the minds of the public, it seems that the deaths of the people at the hands of the terrorist reverberate for years. Is it this public consciousness that makes the public transport systems such an attractive target for terrorists? Ultimately, these types of attack achieve two aims for the terrorist groups: they terrorise the public and they cause substantial economic damage to the infrastructure. Under such circumstances, it is perhaps not surprising to discover that most reviews of terrorist attacks typically find that between 30 per cent to 42 per cent of all terror assaults are directed against mass-transportation systems (Dolnik, 2007). Certainly, there is no sense that the current terrorist interest in striking such targets has diminished.

In an analysis of the Mumbai train attacks in July 2006, which resulted in 209 deaths and over 700 people being injured, Chakravarthi (2006) argues that mass-transit systems are the preferred target of choice for terrorists because they present high concentrations of people and provide ample scope for large scale destruction. Besides facilitating travel and mobility for the people, a nation's economy hugely depends on the road and transit systems. Hence, sabotaging them is akin to killing two birds with one stone – terrorising the population and causing economic damage. The media response to these attacks arguably reinforces public anxiety and helps to create the atmosphere that the terrorist is seeking to achieve.

In 2004, in preparation for the Beijing Olympics, the Chinese government built a significant extension to the public transport network in order to increase daily capacity by four million extra passengers (ABC News, 23 June 2008). For the Athens Olympics, the estimated number of spectators was five million (Loucopoulos and Prekas, 2003). These figures illustrate with stark clarity the importance of public transport systems in getting spectators to and from Olympic venues. Most capital cities experience severe road congestion every day of the week and, given this typical state of normality, it is simply not possible to move large numbers of Olympic spectators around without relying upon mass-transit systems. As noted in earlier chapters, we have seen in many of the terrorist attacks that have taken place across the globe that the current international terrorist methodology is to kill large numbers of people. The very fact that high numbers of people attend major sporting events renders them and busy transport modes an attractive target for terrorist organisations.

The UK's 2000 Terrorism Act defines terrorism as 'the use or threat of violence to intimidate the public or a section of the public' (Walker, 2002). Bearing this in mind, we can see that Olympic events provide an ideal platform for a terrorist group aiming to kill or injure large numbers of people – hence, the substantial resources that are invested in detecting, deterring or disrupting terrorist attacks. Of course, many spectators to Olympic events will travel from outside the host country, so, in examining transport security, we need to assess all the modes of transport that will be used during such an event. As people will be arriving by air, sea or land (either by rail or road), transport security needs to consider all of these modes, some of which will be easier to secure than others.

Road transportation systems

If we consider road transport (either car, bus or coach), the most obvious solution is exclusion – i.e. not allowing any vehicles within the vicinity of an Olympic event. In practice, however, this is not a straightforward remedy. One method of moving significant numbers of spectators around is the use of buses or coaches, usually dedicated for this purpose. At the Athens Olympics, for example, there were dedicated coach and buses (provided free of charge at designated stations) in order to enable the efficient movement of spectators to and from the stadia. These vehicles can take the spectators to within a reasonable distance of the stadium where they can discharge them and pick up returning passengers. Vehicle barriers are also routinely used to prevent any hijacked vehicles penetrating the security cordon. In the early days after the New York attacks of 11 September 2001, concrete barriers were deployed around key areas of London to provide stand off to vehicle-borne attack.

Such barriers, however, are not always effective. The terrorist attacks in Riyadh in 2003, for example, involved the use of multi-wave attacks. The first wave was designed to defeat the defences and enable the terrorists to gain access with explosive devices carried by the following waves. The attack demonstrated the potential weakness of concrete barriers in the face of a determined assault. Subsequent tests by the Home Office Scientific Development Branch (HOSDB), in conjunction with the Centre for the Protection of the National Infrastructure (CPNI) in the UK, demonstrated further weaknesses in these barriers. Detonating an explosive device in close proximity to a concrete barrier destroys the barrier and the fragments of concrete become airborne missiles, adding further to injury and damage to people and buildings. Follow-on experimentation has led to the design and deployment of blast-resistant barriers around iconic locations that are designed to defeat multi-wave terrorist attacks.

The use of such barriers around Olympic venues, with sufficient stand-off, can do much to mitigate the effect should such a methodology be utilised by terrorists. However, there are downsides to the use of barriers around an Olympic venue. When the UK bid to hold the Olympics was being developed, the author was involved in discussing the likely security regime with the UK bidding team. Much emphasis was put on the ideal that the Olympics are first and foremost a sporting event, with security issues coming second. Images of a fortress around the Olympic venues do not fit well with this concept. This issue has also been raised in relation to the Vancouver Winter Games in 2010 by Michael Zelukin (2009: 14) who noted:

> Canadian officials recognize that a reduction in visible security may weaken a deterrence effect, but images portraying a police state with 'barbed wire, armed roadblocks and military carriers in the streets' can be as damaging to the spirit of the Olympics as a backpack bomb in a celebration plaza.

This, then, is the dilemma facing those officials responsible for providing security at an Olympic event – to make the venues friendly and inviting, whilst at the same time trying to keep the venues, athletes, officials and spectators safe from terrorist attack.

Another precautionary measure would be for the drivers to search the buses and coaches before passengers would be permitted to board. This tactic has been deployed in many situations in the UK, with the vehicle being searched by trained staff and then kept secure by the driver throughout the day. The search routine could be undertaken on a daily basis, or the vehicle could be searched at the beginning of the event and kept in a secure location when not in use. These tactics are recognised methods of securing this type of transport.

Without a search of the boarding passengers, which, during peak times, would possibly be too time consuming and require large numbers of trained staff, there is also, therefore, a risk of terrorists boarding a vehicle in possession of an improvised explosive device or firearm. A risk assessment would have to be made of the threat in this type of scenario. Armed law enforcement officers in effective numbers could be deployed to the dedicated bus and coach stations, which should act as a deterrent. Should the would-be terrorist get through and reach the venue by this type of transport, then a search regime at the venue on persons entering should provide the necessary safeguards. Significantly, the quantity of explosive that can be transported on foot is such that it is unlikely that the terrorist would use this method of transport to get to the stadium. There are many documented instances where buses have been targeted by suicide bombers, both in the UK and the Middle East, but the methodology described above would mitigate the damage to the actual Olympic venue. With regard to cars, having excluded the general public and with the judicious deployment of blast-proof vehicle blockers, the only other option would be to purport to be a VIP and gain access through the VIP vehicle system. Again, as described earlier, the searching and securing of VIP vehicles, with an accredited system of passes and effective access control systems, should mitigate against this type of scenario.

Rail transportation systems

If we now turn to rail systems, most major cities around the world have mass-transit rail systems, either under or over ground, in order to transport large numbers of residents, workers, visitors and tourists around the city. The volume of passengers transported on such systems in London is formidable: over one billion passengers were carried on London's underground system in 2008/2009, averaging over three million passengers per day (Transport for London, 2009). Given the massive numbers involved, the system's ticketing processes have been designed to ensure rapid access through the turnstiles to reduce congestion and prevent the build-up of large queues. Even with these fast-flowing systems, during the peak periods in the morning and evening there can still be significant queues of passengers waiting to get through the turnstiles onto the platforms. The introduction of the card based 'Oyster' system has increased the flow rates even further, but passengers can still be subject to delays during such periods.

Following the terrorist attacks on the London Underground in July 2005, there was some media discussion in relation to securing mass transit systems in the way that is practised at airports. However, an examination of the volumes of traffic and the geography of the two systems reveal very different situations. According to the figures from the British Airports Authority (2010), which is responsible for the maintenance and operation of Heathrow Airport, there are something approaching 186,000 passengers passing through the five terminals daily at the airport. There is a very clear line of demarcation between the 'landside' and 'airside' areas of an airport, where passengers have to show a boarding pass and go through a search regime. Whilst this is an acceptable necessity with the numbers at an airport, amounting to around 6 per cent of the numbers that use the underground system, any similar system for the latter would be cost-prohibitive and grind the network to a halt very quickly. Also, passengers travelling on an aircraft are, by virtue of the publicity, 'trained' not to carry unsuitable articles on their person. The prospect of trying to persuade workmen travelling on the underground not to carry sharpened instruments, for example, would prevent them from being able to go about their business.

During a visit by the author to Russia to examine the bombing of the Moscow Metro system in February 2004, a system of recording passengers travelling on the national railway system was demonstrated to the UK delegation. Every passenger travelling on this network has to show their identity card when purchasing a ticket for travel and they are allocated a carriage and seat number. Their details are entered into a national criminal database and wanted persons are immediately identified, with officers despatched from the local police station to apprehend them – at first review, a very effective system. However, the Russian authorities informed the author that there had been considerable difficulties in passing the necessary legislation through the Duma in order to enforce this regime. Also, this type of legislation would be a considerable intrusion upon the private lives of citizens and, given the challenges faced by the Russian legislators (even in a country with a suspect human rights record), passing similar legislation in a Western country would be particularly difficult, perhaps exemplified in the debate in the UK around the controversial proposal for identity cards. Moreover, if the would-be terrorists were 'clean skins', as was effectively the case with the London and 9/11 bombers, then checks of this type would not identify them as would-be terrorists. So what we are left with is attempting to manage very large numbers of persons travelling on the railway network, largely unimpeded, whilst trying to identify the miniscule number who would seek to cause harm.

In the summer of 2006, there were a series of trials conducted by the Department for Transport, in conjunction with the British Transport Police and Metropolitan Police on the over and underground network following the July 2005 terrorist attacks (utilising existing technology, archway metal detectors, X-ray and vapour machines, and explosive detector dogs) in order to determine their acceptability to the travelling public. There was no attempt to measure their effectiveness against potential terrorists as blind trials with persons carrying explosives would have had to be introduced. There was a realisation that the percentage of persons screened would be so low as to render the trial ineffective. Instead, the trial was much more about testing the travelling public's reaction to the possibility of introducing such a security regime (Department for Transport, 2008). The main outcomes from this research were twofold. First, that the current security scanner technology was not yet sophisticated enough to provide an effective search regime for travellers on the rail system and, second, likely delays, issues of privacy and the selection process all gave respondents cause for concern over the use of such technology.

One of the technologies being studied, but which is not yet anywhere near the stage where it could be deployed, would be one that would enable the detection of chemicals on a person's skin as a consequence of handling explosives or the precursor chemicals. It is a well known phenomenon that exposure to explosives can be detected through the sweat of persons who have handled them (National Research Council, 2004). There are technologies in embryonic form that may be able to do this. The ideal solution would be a ticket machine that could 'sniff' the ticket as it passes through the ticket reading machine to detect the illicit substances. The growing use of the 'Oyster' technology, however, works against this type of detector, particularly as efforts are underway to develop proximity machines to read tickets without having to hold them against the reader.

Even if it was straightforward to implement – which is not currently the case – there are two further drawbacks for these types of technology. The first one would be the large number of false alarm calls, as many materials may trigger an alert from the machine. In time, however, it is probable that developments in the technology would significantly reduce the false alarm rate. The second problem would be the speed and type of response from the law enforcement agencies. During a global research project into the phenomenon of suicide terrorism, the author was informed of many examples where suicide terrorists had prematurely detonated their devices because they thought they had been detected. This was very common in Russia and Israel. One of the Russian examples included two Chechen female suicide bombers on their way to attack a stadium when they spotted the police carrying out searches. They left and detonated their devices a short time afterwards. In Israel during the height of the terrorist campaign, there were numerous examples where a security guard had stopped a suspected suicide bomber who then detonated their device, killing both themselves and the security guard. Unless a disproportionate number of armed officers were deployed on the network, the response would be measured in tens of minutes. In this time the suspect could board the train and be lost in the system.

The alternative would be to have a bomb proof portal that passengers would walk through when presenting their ticket. The portal would have to be designed to allow one person at a time to transit and an alert would automatically enclose the ticket holder pending arrival of the police. The potential negative consequences of this could be the person trapped actually being innocent and being subject to false imprisonment, or a terrorist detonating the device and potentially injuring passers by, depending upon the strength of the enclosure. Another scenario in this situation is the possibility of law enforcement officers facing a bomber with a live explosive device. How would they deal with such a situation? There would be no guarantee that the device was safe if they tried to detain the person even after the bomber may have apparently disarmed him or herself. It could be argued that in these circumstances the police would be justified in using lethal force under Section 3 of the Criminal Law Act, 1967, but, having already been imprisoned by the technology, there is the question of the suspect's human rights.

Aviation transportation systems

Another key challenge in securing transportation to an Olympic event is to protect against the potential for an attack on the aviation transport system. As discussed earlier, aviation has proved to be an attractive target for terrorists, whether it be to use the aircraft as a weapon to damage infrastructure, or to hijack an aircraft and take hostages in order to extract concessions. Significant measures have been implemented to secure aircraft from attack, from reinforcing the cockpit doors, to preventing the terrorists seizing control, to strict searching regimes and baggage checks prior to boarding. These measures have proved to be largely successful. However, if we examine some of the incidents that have taken place since the 9/11 attacks, then we can see that terrorists are still searching for ways to defeat the security regimes. On 22 December 2001, for example, just three months after the 11 September attacks, Richard Reid managed to board an American Airlines flight with a bomb concealed in his shoes. It was only his own ineptitude in trying to detonate the device that alerted the cabin crew and passengers to his activities and he was successfully restrained.

The liquid explosives plot of August 2006 (that was foiled by the UK authorities) is another example where terrorists devised a scheme to try to circumvent aviation security measures by smuggling liquid explosives onto an aircraft disguised as soft drinks. The authorities are always at the mercy of the ingenuity of the terrorist and constantly need to be alert to these inventive schemes.

Behavioural studies is an area that is receiving a significant amount of attention and research at the present time (Silke, 2010). Some characteristics of this, and suggestions about how to exploit it, are described later in the chapter, together with some interesting learning points that are emerging from the research into previous attacks and attempted ones. The threats to the aviation industry from terrorism have, over the years, produced an iterative effect on security regimes for passenger terminals at airports and the screening systems for departing travellers.

The ideal situation with regard to security at an airport would see everyone who had cause to enter an airport terminal being searched and cleared, either before they entered the terminal, or immediately after they have entered the terminal building. This would create a situation whereby the whole terminal could be considered to be a sterile area. Therefore, anyone wishing to cause harm to the travelling public or damage to a terminal building could be intercepted by an armed response unit before they were able to prosecute their mission.

Indeed, this principle is adopted in many building locations around the globe, where visitors are checked and cleared before they enter a building. However, for an airport terminal building, this is not practicable. Departing passengers are often accompanied by friends or relatives seeing them off and, in a similar manner, arriving passengers are often met at the gate by their friends or relatives. Added to this are taxi drivers picking up visitors, people visiting the airport for information or purchasing tickets, or even people wishing to visit an airport out of pure curiosity. The demands placed upon a screening regime in such circumstances would be impossible to satisfy and the costs would be prohibitive. Therefore, the screening regime has to take place at a location where the travelling public can be separated from the casual visitor in a controlled area. In some airports that the author has been through, passenger screening does not take place until the passenger reaches the departure gate, although in the majority of airports the screening typically takes place at the landside/airside interface, where different regulatory regimes can be enforced. As a consequence, aviation security is usually an amalgam of tactics, and these would usually seek to eliminate the opportunities available to the terrorist. For example, some of the tactics would mitigate against the threat to the landside areas of the terminal, by the use of regular searches to identify suspect devices left in these areas, with armed law enforcement or military patrols to deter the would-be attacker. The main tactic to prevent attacks in the airside areas of a terminal is to screen all persons passing into these sterile areas, be they departing passengers or employees there.

Some of the other factors that need to be considered in securing the system is the role of airlines, ticketing procedures and the profiling of passengers. The bomb that brought down the Pan Am 747 flight in December 1988 over Lockerbie was placed in a container at Frankfurt airport and transferred to another container at Heathrow where it was placed on the ill-fated aircraft, which subsequently exploded with great loss of life. Even though the perpetrator(s) defeated the security regimes at Frankfurt and Heathrow, both airports are still operating and thriving. Pan American went out of business as a consequence of the bombing, and yet airlines continue to abdicate their responsibility for security and leave it in the hands of the airport operators, even though the airline would suffer far more than the airport were an incident to take place. The attempted bombing on 25 December 2009 of the Northwest flight 253 from Amsterdam to Detroit is a clear example of where the airlines are culpable in this regard. Umar Farouk Abdul Mutallab was travelling alone, bought a one-way ticket with cash and was carrying no hold baggage, only cabin baggage (a strikingly similar profile to that of Richard Reid in December 2001). This information was known to the airline before he arrived to check-in. These pieces of information alone should have precipitated a more thorough scrutiny of his reason for boarding this flight. The airport operator, who is charged with carrying out the search regime, could not have known this information and was therefore starting from a zero knowledge base. Had the airline passed this information to the airport operator, he could have been subject to a more rigorous search regime which should have found the concealed device. There also appears to be a further intelligence failure given there was awareness in some quarters regarding his history of radicalisation, but this does not excuse the airline from carrying out a more detailed investigation prior to his arrival, purely on the information that they were in possession of about his ticket purchase.

If we look, however, at the use of technology to book flights, there is much that airlines could do to increase their own security as part of a partnership between themselves and the airport operators. The key to this is the 'profiling' of passengers, a 'dirty' word in many people's eyes. This is because profiling, amongst the less sophisticated techniques, tends to use the colour of a person's skin as the profile differentiator. However, if we look at the air passenger, we can probably start profiling by drawing a distinction between the business traveller and the leisure traveller. Business travellers tend to book their tickets through a travel agent used by their parent company, their tickets are paid for by invoice or company credit arrangements, and they are well-known to the industry. They are very likely to have a frequent flier card and when they turn up at the airport to check in, the airline transporting them already knows a significant amount about that person. A differentiated boarding ticket could be issued, so that when going through the search regime they could be subject to a lighter touch, if any security process at all.

Leisure travellers, on the other hand, may be flying for the first time. They may have paid cash or used some other means for the purchase of the ticket and would be relatively unknown. Such passengers should be subject to the full search regime. Taking the business out of the search process gives more capacity to focus on the leisure flyer and everyone benefits from less inconvenience and faster transit times. There will always be exceptions to this differentiation, but the situation with current security regimes is that terrorists have succeeded in significantly driving up the costs of security as well as causing delays, to the detriment of the industry and the travelling public.

Every new threat brings yet greater levels of security, without any real review of what the risks are and what is a proportionate response. There are clear opportunities to learn from the various successful and unsuccessful attacks that have taken place around the globe that have been aimed at the aviation industry. However, the stock answer by the regulatory authorities has been to introduce ever more intrusive measures, such as the banning of the carriage of liquids, leading to longer queues and an altogether unpleasant experience for passengers.

Marine transportation systems

If we now turn to the issue of the security of water-borne transport and its potential to be a target, there are a number of considerations. First, and probably of least concern, will be spectators arriving in the UK by water, generally a ferry or (less likely) a cruise ship. The ports are generally well protected owing to the measures that have been put in place to detect illegal immigration, or the smuggling of human beings, drugs and other items of contraband into the country by organised gangs. All persons arriving into the UK, be they nationals returning home or visitors, have to go through passport control where their passports are scanned to ensure they are valid and that they match the person presenting them. Those travellers requiring visas also have the validity of these checked. Once through passport control, entrants have to pass through a Customs check where stop-and-search procedures can be carried out to check for illegal possession and importation of drugs, contraband substances and trafficked human beings.

One potential scenario is that a ship could be attacked and hijacked by terrorists as a means of transportation. Past examples of this approach include the *Achille Lauro*, which was hijacked in 1985, and the trawler that the Mumbai attackers of November 2007 hijacked in order to reach Mumbai. However, given the geography of the UK, ships seized in such circumstances would reach the coastline but would be hard pressed to penetrate far inland without the authorities noticing. One interesting aspect of the seizing of ships is the issue of powers to board and search vessels. The territorial waters extend to a distance of 12 miles from the shoreline and, once within this distance, the home authorities have powers to engage and board such vessels. However, law enforcement or military authorities wishing to board a vessel outside their territorial waters, if that vessel is not registered to the home country, have no powers to board except with the express permission of the crew or the state where the vessel is registered. In the event of a biological weapon being carried by terrorists, with the right weather conditions, a weapon of mass destruction could therefore be discharged from outside the 12-mile limit and the substance could still reach the mainland.

Responses to such scenarios have been exercised in the UK, with much debate amongst the legal teams involved. The consensus reached, however, was that if there was reasonable cause to suspect a vessel was carrying this type of device, then action would be taken and the legal issues resolved subsequent to the intervention. This is an area that needs further discussion internationally. The location of the London Olympics does pose a particular challenge for the security authorities owing to the waterways that run around and through the Olympic sites. Spectators attending the events may well be transported to the venue by water and measures will have to be put in place to ensure that any possibility of inserting a device is deterred or detected. The other threat posed by the waterways is penetration by terrorists utilising diving equipment. Work will have to be undertaken to secure this method of entry, either through the deployment of technology, or regular patrolling, or a combination of the two.

Suicide terrorism and the transport system: opportunities for interdiction?

The events of 11 September 2001 have precipitated a huge programme of research into new technologies, policing tactics and protective security regimes to counter the threat from international terrorists. One of the most difficult challenges arising from contemporary international terrorism is the threat from suicide bombers. Eliminating the need for an escape plan renders many of the counter-terrorist techniques developed during campaigns such as that waged by PIRA ineffective (Drake, 1998). As a consequence, millions of pounds have been expended in trying to counter this particular threat, but with limited success. However, the attacks in London in July 2005, and those seen in other places where suicide bombers have been deployed, have created a database of sufficient size that it is possible to examine the characteristics of the persons involved in these attacks to look for predictable behavioural patterns that may permit some exploitation by law enforcement agencies, particularly when it comes to preventing attacks on transport systems.

The current methodologies attributed to Al Qaeda and its affiliated groups are mass casualty terrorist attacks, and these have occurred in many countries around the world. Yet, in many cases, the death toll is much lower than it could have been, given the availability of 'soft targets' and the apparent ease with which these terrorists seem able to penetrate protective regimes. To the outside observer this seems illogical, particularly given the inherent flexibility of a terrorist embarking on a suicide mission and his/her ability to change targets and locations right up to the point of detonation. Investigations into these terrorist attacks reveal clear evidence of reconnaissance of the target prior to the attack. This reconnaissance should ensure a level of knowledge, planning and awareness that gives them the best times, locations and targets to meet their mission objectives. In a study of over 300 suicide attacks, Pape (2005) found that the average number of deaths per attack is 12 (excluding 9/11). However, many attacks fail to meet this 'average' impact and some useful insights can be drawn from examining some attacks which arguably could have been much more severe.

On 7 July 2005, at 08.50, three of the four bombers that attacked the London transport system that day, Tanweer, Lindsay and Khan, carried rucksacks onto the London Underground railway and detonated them, killing and injuring many travellers. The trains at Aldgate and Edgware Road were crowded but not full. However, the King's Cross Underground train was at or near full capacity. The devices were placed on or near the floor in the standing area next to double passenger doors. The device at King's Cross was more centrally placed than at the other two Underground incidents. A total of 27 people died at the King's Cross incident, approximately four-times as many deaths as in either the Aldgate or Edgware Road incidents. The high number of fatalities at King's Cross is likely to have been caused by the central location of the device and the passenger distribution. The small bore of the tunnel is not considered to have been a major factor in this instance.

Nearly an hour later, at 09.47, the fourth bomber, Hussain, detonated a similar bomb on the upper deck of a London bus, again killing and injuring many people. Owing to the death of the subject, some of the following information can only be surmised. However, given the almost simultaneous detonation times of the three bombers on the Tube, it seems likely that Hussain was supposed to detonate at the same time. CCTV recordings reveal that the bombers had carried out reconnaissance activity some nine days previously (Intelligence and Security Committee, 2006). However, on 7 July, Hussain was unable to implement his pre-planned attack owing to disruption of the Northern line. He is seen on CCTV wandering around for nearly an hour before boarding a bus and completing his mission. During this time, he tried to phone his fellow terrorists, when he must have known that they were already dead. Some 14 people died in the Tavistock Square bus attack, at least eight of whom were believed to have been on the upper deck where the device was detonated.

On 21 July 2005, exactly two weeks after the 7/7 attacks, a fresh group of terrorists attempted to carry out another wave of suicide bombings on the Underground system and buses in London. At 13.07 hours, one of these new attackers, Muktar Said Ibrahim, can be seen on CCTV taken from a London bus on which he was travelling, sitting in the back of the bus attempting to detonate the device. The moment of detonation can be seen as the passengers look back in his direction. At this point, there are five other people on the top deck, one on either side of the bus three rows ahead, one halfway down and two right at the front. Had the bomb detonated, given the distance between Said and the other passengers, and the small numbers of them, it is likely that his would have been the only death, though, undoubtedly, others would have been injured.

The author has also seen some footage from a vehicle checkpoint in Iraq, where a vehicle-borne suicide bomber detonates his device, but only kills the soldier standing next to the vehicle. In this attack, the driver of the VBIED can be seen approaching the checkpoint before stopping some 50 yards away. After a very short period of time, the vehicle then approaches the VCP. At this point there are six vehicles waiting at the three lanes of the checkpoint, one of which is a US military vehicle. The terrorist vehicle draws up to the rear of the queue and then moves forward until it reaches the head of the queue. A soldier leans forward to speak to the driver and then, as he steps back, the device detonates. At this point there are no other vehicles at the VCP so the only person apparently killed by the blast is the lone soldier. In all these examples, the bomber(s) has failed to maximise the effectiveness of his attack. Something seems to be occurring in relation to the behaviour of the bombers in these examples. What is it? Is it possible to draw any hypotheses that can be utilised by law enforcement agencies?

Silke (2006) conducted a study on psychological aspects of actually carrying out a suicide attack. One of the conclusions from this was that suicide terrorists typically experience very high levels of stress when they are attempting to execute the attack. This stress, Silke argued, has a very clear impact on the bomber's ability to think clearly and react to situations:

> In high stress situations, individuals have difficulty concentrating, display poorer memory, and have greater difficulties with problem solving and perception tasks. The result is that decision making in such circumstances can take longer and is much more prone to errors and poor judgement. Training and preparation to prepare for stressful circumstances, and previous experience of similar situations, have been found to help reduce these negative effects (though they do not entirely eliminate the impact).... Suicide terrorists often receive relatively little training and preparation.... The result is that suicide terrorists may have difficulty effectively responding to unexpected obstacles or changes in plan. Unanticipated events will cause terror-

ists to take longer to reach decisions and the decisions made are more likely to be poor ones. Even minor unanticipated disruptions to a plan can have a considerable impact in terms of the outcome (e.g. a bomber detonates in an area with a poor potential for mass casualties; or, abandons the attempt entirely). Such a relationship endorses security approaches that increase the likelihood of unanticipated changes (e.g. random security checks at travel hubs, etc.).

(Silke, 2006: 3)

There is another interesting phenomenon that has emerged from an examination of the facts surrounding these attacks, which relate to the geography of the underground system, and was seen in both the Moscow Metro bombing on 4 February 2004, the London bombings of 7 July 2005 and the failed attacks in London on 21 July 2005. When the attack on the Moscow Metro was investigated, an interesting fact emerged during a visit to the scene. Tracing the route of the bomber to the position where the device detonated revealed that the bomber had taken the shortest route possible from the surface to the platform. At the bottom of the escalator the bomber turned immediately right and boarded the carriage at this point, which is where the device was subsequently detonated. Upon examining the London Underground explosions of 7 July and the failed explosions on 21 July, it can be seen that four of the bombers were in the second carriage of the six car train, one was in the first car and one was in the fourth car. An obvious question is, why choose these particular locations? When their routes are retraced (and some of this has to be supposition, as many of the CCTV systems were not working), it is possible to give a logical explanation for this. For the 7 July attacks, walking the shortest route from King's Cross mainline station through the underground tunnel system brings each of the three bombers onto the platform opposite the carriage from where they boarded the train and then subsequently detonated their respective devices. This same fact is repeated on the 21 July example. In each case, the bomber appears to have entered the carriage closest to where they first emerged on the platform. Once they arrive on the platforms, they have remained static.

By and large, the intention of most suicide bombers is to cause as many casualties as possible. They undertake reconnaissance of their intended target in order to identify vulnerabilities and weaknesses, so they can plan and execute their attack. During the time when they are prosecuting their attack, the extreme stress faced by them during the period leading up to the point of detonation is interfering with their cognitive abilities and their mind set is preventing flexibility (Silke, 2006). The way that security processes generally work means that the regimes deployed are quite predictable, as they need to be managed to ensure their effectiveness. This predictability actually facilitates the planning processes undertaken by the terrorist and reduces the effect of the extreme stress. Also, current security regimes are very open and transparent. Generally there will be archway metal detectors and baggage X-ray machines. Some locations may have vapour technology machines where the operator will swab the visitor's bag to check for traces of explosives. Therefore, the terrorist, when carrying out reconnaissance on their potential targets, can plan an attack having been able to witness first-hand the existing security regime.

Following from this, the obvious question for law enforcement agencies is: what can we do to exploit this behavioural anomaly, as even small changes in routine may deter, detect or disrupt them? This is often easier said than done, as infrastructure changes cannot be undertaken easily. Nevertheless, operators of transport systems should review their security regimes to look at how they can make their processes less predictable and more opaque. One way may be to move the emphasis away from having highly visible guards to some who operate in non-descript clothing, perhaps working in the area in front of the location to be protected, so they are able to observe behaviours and perhaps call attention to anomalous behaviour. Some of the solutions may be: shifting entrances to places of public resort if possible; closing entrances for short periods of time; introducing security guards at irregular intervals; focusing activity at the points of shortest route. Other agencies may have ideas about changing routines. There is obviously no guarantee of success as the terrorist's reconnaissance may coincide with changes. Also, this is only likely to delay the terrorist and not to stop him or her. However, delay may give law enforcement agencies an opportunity to spot unusual behaviour or cause the terrorist to act sufficiently outside the normal patterns of behaviour as to enable the authorities to spot and intercept them, particularly if staff within selected organisations are briefed on the phenomenon. The author has discussed the feasibility of introducing some of these ideas into security regimes in the aviation transport environment. However, owing to the regulatory nature of aviation security, little progress has been made. The 25 December 2009 attempted bombing discussed earlier in this chapter may well prove to be the catalyst for a change of thinking about such suggestions. Some of this unpredictability may, however, also cause some difficulties for the travelling public. Notwithstanding this inconvenience, changing the routine of security procedures may be the price to pay for making the transportation systems safer for the travelling public.

Conclusion

Public transport systems will remain a favourite target of terrorist groups because of the two goals such attacks achieve: the terrorising of the public and significant economic damage. Ultimately, all major public transportation systems, whether by land, air or sea, are vulnerable to attack, regardless of the substantial resources that have been put into developing technologies to disrupt, deter or detect the terrorist. Much work is needed to develop new and innovative technologies to identify and apprehend the terrorist, particularly at major sporting events such as the Olympics – because of the large numbers of spectators and the current difficulties in providing cost-efficient and effective preventive measures. Technology alone, however, should never be seen as the only solution to terrorist threats. Other issues are also vital. In the context of the Olympics, for example, the training of the large body of volunteers and security staff to perform the security function and use the security equipment is critical because the technology will only ever be as good as the person operating it.

One of the drawbacks of security is the significant cost involved, which is often seen as a 'straight off the bottom line' cost. Consequently, in the desire to cut costs, often relatively unskilled staff are employed to carry out these roles, particularly if the period of employment is temporary, as would be the case for a sporting event such as the Olympics. This means that the training regimes for the staff have to be standardised to enable them to perform the role reasonably effectively. Standardised training is predictable by its very nature. Therefore, for the terrorist planning an attack, they are able to factor in this predictability when developing their attack plans. The challenge for the security designers and planners is to build unpredictability and opaqueness into the Olympic security regime and all the associated transportation systems, so that planning an attack becomes much more difficult and gives the law enforcement agencies opportunities at the intervention points to spot danger and to take action.

References

ABC News (2008). 'Beijing to Launch Olympic "odd–even" car ban.' www.abc.net.au/news/stories/2008/06/23/2282484.htm?site=olympics/2008.

British Airports Authority (2010). 'February Traffic Figures: BAA's Airports.' www.baa.com/portal/page/BAA%20Airports%5EMedia%20centre%5ENews%20releases%5EResults/915008f679e27210VgnVCM10000036821c0a____/a22889d8759a0010VgnVCM200000357e120a____/.

Chakravarthi, R. (2006). *Mumbai Train Attacks: Why Do Terrorists Target Public Transport?* www.ipcs.org/article/india/mumbai-train-attacks-why-do-terrorists-target-public-transport-systems-2070.html.

Department for Transport (2008). *Transport Security Measures – Attitudes and Acceptability.* www.dft.gov.uk/pgr/security/sectionsocresearch/security/.

Dolnik, A. (2007). 'Assessing the Terrorist Threat to Singapore's Land Transportation Infrastructure.' *Journal of Homeland Security and Emergency Management*, 4/2, Article 4.

Drake, C.J.M. (1998). *Terrorist Target Selection.* New York: St Martin's Press.

Intelligence and Security Committee (2006). *Report into the London Terrorist Attacks on 7 July 2005.* London: HMSO.

Loucopoulos, P. and Prekas, N. (2003). 'Designing Venue Operations for the Athens 2004 Olympic Games.' *The Connector*, 1/2. www.iseesystems.com/community/connector/zine/march_2003/Prekas/index.html.

National Commission on Terrorism (2000). *Countering The Changing Threat of International Terrorism.* www.fas.org/irp/threat/commission.html.

National Research Council (2004). *Existing and Potential Standoff Explosives Detection Techniques.* Washington, DC: National Academies Press.

Office for National Statistics (2009). *Mortality Statistics: Deaths Registered in 2008.* www.statistics.gov.uk/downloads/theme_health/DR2008/DR_08.pdf.

Pape, R. (2005). *Dying to Win: the Strategic Logic of Suicide Terrorism.* New York: Random House.

Silke, A. (2006). *The Impact of Stress and Danger on Terrorist Planning & Decision-Making*. London: Andrew Silke.

Silke, A. (2011). *The Psychology of Counter-Terrorism*. London: Routledge.

Transport for London (2009). *London Underground Named Best Metro in Europe as Passenger Numbers Hit Record High.* www.tfl.gov.uk/corporate/media/newscentre/archive/11579.aspx.

Walker, C. (2002). *Blackstone's Guide to the Anti-Terrorism Legislation*. Oxford: Oxford University Press.

Zelukin, M. (2009). 'Olympic Security: Assessing the Risk of Terrorism at the 2010 Vancouver Winter Games.' *Journal of Military and Strategic Studies*, 12, 1, 1–25.

6 Surveillance and the Olympic spectacle

Pete Fussey

Since Munich, one of the key features of Olympic security has been the use of technological surveillance, a strategy that has become increasingly central to securing large sporting events in the post-9/11 era and one that fits neatly with the IOC's demands to prioritise the sporting event over the policing spectacle. With the London Games likely to become the first biometric and wireless Olympics, the capital is likely to reinforce its reputation as a pioneer of such technologies via the deployment of ever-more intensified, networked and advanced forms of technological observation.

This chapter examines the way technological surveillance has been applied to secure mega sporting events (with particular reference to the post-Munich Olympiads) and considers the implications of these processes and practices for 2012. In doing so, key processes shaping the form and scale of surveillance strategies and the types of technological surveillance provision (from first generation CCTV systems to second generation video analytics) are identified and analysed before their application, impact, efficacy and legacy are critically assessed.

As this chapter argues, a broad view of Olympic security operations reveals the convergence of at least three different processes. First, a number of security themes have been transferred across events and locations. Second, their direction and intensity have also been shaped by responses to key events, most notably, Munich, the 1996 Atlanta bombing and 9/11. At the same time, threats to Olympics have constantly shifted in relation to their own contextual environments and logic, thus raising operational questions over the relationship between future planning and retrospective events.

In exploring the role and impact of surveillance strategies in securing the Olympics from crime and terrorism, this chapter engages in three main areas of discussion. First, a number of key contextual issues are addressed. This involves a definition of what is meant by 'surveillance' and an exploration of surveillance strategies in the areas hosting the 2012 Games (both nationally and locally). Second, the chapter will focus on the ways sporting events, particularly the Olympic Games, have been secured using surveillance technologies. Finally, the chapter will draw on some of the key analytical issues emerging from these discussions to consider the efficacy, operational context and ethical considerations of surveillance strategies at Olympic sized events.

Surveillance in context

Defining surveillance technology and its application

This discussion will first briefly explore what is meant by surveillance in this context. One of the appeals of surveillance strategies for practitioners is the view that it can undertake myriad functions and segue with strategies already operating to tackle crime and terrorism (Fussey, 2007b). Although surveillance strategies may take 'informal' and 'low tech' forms – such as increasing pedestrian flows, enhancing sight-lines by removing obstacles such as hedges and the adoption of more open plan architectural techniques – the most recognisable and debated form of surveillance has been closed-circuit television (CCTV) cameras that observe urban spaces and feed footage back into a control room. Yet, as numerous studies have illustrated, this form of observation is far from passive, nor does it operate in a linear directional manner. For example, ethnographic studies have demonstrated how the subjects of CCTV observations reflexively amend their behaviour in response to their perceived visibility – becoming *active* 'subjects of communication' rather than *passive* 'objects of information' (Smith, 2007: 280). The surveillance technology itself has also become more active over recent years. As discussed in more detail below, the past decade has seen the growth of second generation surveillance technologies designed to organise, filter or elevate noteworthy signals from the white noise of oversupplied surveillance data.

Yet technological surveillance also involves more than observing the bodies of suspects. Increasingly, technologies have moved towards the surveillance of *mobilities*, from long-standing remote sensing techniques (for example, via infra-red or ground sensors) to mobile phone signal triangulation and GPS techniques (including the recent satellite tracking of repeat offenders, see Shute, 2007), to recent attempts to monitor the movements of spectators of sporting events by embedding Radio Frequency Identification (RFID) chips within their tickets.

Connecting with Mathieson's (1997) concept of the 'synopticon', where the many watch the few (rather than the 'panopticon' where the few watch the many, see Foucault, 1977), such technologies have also been deployed to monitor the activities of security agents. Although some examples of this inversion of the surveillant gaze have been punitive – as are apparent in the infamous broadcast of Rodney King's beating at the hands of LAPD officers[1] and the more recent suspension of a Greater Manchester Police officer following violence against an individual suspected of damaging a bus shelter (Carter, 2004) – it is more frequently deployed to enhance strategic operations. For example, a rationale for one of the first Olympic CCTV deployments, at Montreal, was to help coordinate the unprecedented security operation (COJO, 1976). Since then, similar surveillance applications to improve the coordination and communication between security providers at major sporting events have grown enormously in scale and scope, if not always in efficacy.

Together, these variegated applications of surveillance technologies raise numerous practical and theoretical issues, including questions over their effectiveness, adaptability to new roles (such as Olympic security) and how they shape and are shaped by their social and operational environments. These important considerations are examined later in the chapter in the context of surveillance and major sporting events. Before doing so, however, this chapter will first look at the rise and diffusion of surveillance technologies across both the national context and, more specifically, within the East London boroughs that will host the majority of Olympic events during 2012.

Surveillance and the 2012 hosts – national and local contexts

For a country often accorded the unprepossessing status as the most surveilled democracy in the world, it is perhaps unsurprising that crime related surveillance technology has a long history in Britain. In his illuminating history of CCTV deployment in the UK, Williams (2003) notes the use of Glasgow's camera obscura to arrest a pickpocket as long ago as 1824. However, more recognisable forms of this, such as CCTV, have been permanently deployed in London's public spaces (particularly on Whitehall and around Westminster) since the late 1960s (ibid.). Owing to the then prohibitive costs of cabling and the technological difficulties in ensuring reliable transmissions, here they largely remained until the 1980s.

These early intermittent and largely ill-fated experiments were comprehensively eclipsed by the meteoric rise of CCTV deployment across Britain's public spaces during the early to mid 1990s. From a modest coverage of only three town and city centres in 1990 (NACRO, 2002), only three such spaces remained unobserved by 1996 (Graham, 1998). This 'era of uptake' (Webster, 2004) had stimulated a growth that by the end of the millennium had, according to some, rendered British citizens the most surveilled population in the world (Norris and Armstrong, 1999, although see discussion on the Beijing Olympics below, pp. 103–105). Owing to considerable conceptual and methodological obstacles, the actual number of cameras is unknown, and predictive (and often speculative) estimates range from an unlikely 40,000 (NACRO, 2002) to 4,285,000 (McCahill and Norris, 2004), one for every 14 people.[2] At the epicentre of this network is the capital where, even a decade ago, economically active Londoners could expect to be filmed by an average of 300 cameras a day (Norris and Armstrong, 1999).

The reasons for this meteoric expansion have been varied and have, as Fyfe (2004) correctly notes, tended to withdraw into conceptually utopian or dystopian trenches. For its supporters, CCTV has been regularly portrayed as a necessary tool to address (what were) escalating rates of criminality during the 1990s (see, for example, Horne, 1996). More dystopian accounts – some drawing on the ready (and over simplistically applied) imagery of Orwell's *magnum opus* – have sought to examine the social costs of these technologies. Amongst these, the most robust have focused on the technology once it has become operational, implicating CCTV in the sorting of gendered, racialised or socio-economic 'difference' in late-modern urban spaces (see, for example, Seabrook and Wattis, 2001; Norris and Armstrong, 1999; Davis, 1998, respectively. Also see Lyon, 2003b, for a comprehensive overview of surveillance and 'sorting').

The truth behind the expansion of CCTV, however, is probably more prosaic. Owing to its expense, it has been argued that large scale government funding has been the principle driver behind CCTV expansion (Fussey, 2007b). During the 1990s, government enthusiasm for the cameras was matched by a willingness to fund its deployment under the inaugural 'CCTV Challenge Scheme'. This fervour was maintained by the New Labour government via the Home Office CCTV Initiative during 1999 that allocated £153 million to install public space surveillance systems (see ibid.). Such government spending meant CCTV accounted for over 75 per cent of the Home Office crime prevention budget between 1996–2000 (Ditton, 1999; Wilson, 2002) and led some commentators to estimate that over £500m of public money has been spent on its installation (Norris, 2006). This commitment to CCTV presents two important issues worthy of note. First, this expenditure occurred in the absence of Home Office funded evaluations that proved its ability in the ways many of the tendered bids claimed (see below). Second, this level of government funding has not been matched in many other countries and probably constitutes the main reason for their lower densities of CCTV coverage.

After Belfast, East London (particularly the boroughs accommodating the Olympic spectacle) has, in likelihood, been the site of most experimentation with surveillance in the UK. Long associated as a metaphor for crime and deviance, East London's unique history has been shaped by an enduring and predominant working class that has thrived without the dominant centripetal industries that define blue-collar cultures in many cities of the midlands and the north. For many commentators, this has elevated the need for mercantile dexterity as a strategy for economic survival – an 'entrepreneurial proletarianism' based on the 'commodification of reality' (Hobbs, 1988: 116). As such, the East End has consistently attracted epithets of danger and criminality, serving as a 'writhing symbol of respectable fears and anxieties' comprising a population that is perennially 'dehumanised and classified as deviant by definers whose power base was in the City' (ibid.: 108). At the same time, chronic social and economic disadvantage coupled with perennial demographic transitions (hindering the development of cohesion and established communities) have generated a plurality of criminogenic contexts.

In turn, these (real and imagined) interpretations of 'dangerousness' led to a range of new technologies of control becoming trialled on East London's inhabitants. Examples of this technology include relatively straightforward applications such as Automatic Number Plate Recognition (APNR), developed in the City of London to tackle PIRA terrorism (now deployed throughout East London with particular concentrations in Hackney) and Hackney's 'Shoreditch Bridge' (where local residents could tune into digital CCTV feeds via their televisions). More complex (and normally more unreliable) technologies include one of the world's first public space deployments of Facial Recognition CCTV (FRCCTV) in Newham, host to over 60 per cent of the 2012 Olympic infrastructure, and software that attempts to model human behaviour (such as London Underground's 'Intelligent Passenger Surveillance' technology) trialled at Mile End (in the Olympic borough of Tower Hamlets) and Liverpool Street stations.

Surveillance and the Olympic spectacle

As noted throughout this book, Olympic security is a sizeable and often unwieldy endeavour. In attempting to make sense of this, Sanan's (1996b) discussion of Olympic security between 1972–1996 helpfully defines a number of distinct phases that these operations can be coalesced into. These are: the 'reaction phase' (1976–1980) covering the militarised approach to security immediately succeeding Munich (particularly at Montreal and Lake Placid); the 'hiatus phase' (1980) describing the insularity of Moscow's security planning;[3] the 'consolidation phase' (1984–1988) denoting the internationalisation and integration of security alongside the IOC's enhanced role in this area evident at the Sarajevo and Seoul Olympics; and the 'European phase' (1992–1994) denoting the unprecedented international coordination surrounding the geographically proximate Albertville and Barcelona Games and, also, the 1994 Winter Games in Lillehammer.

Whilst Sanan's heuristic is undeniably valuable, the analysis is not recent enough to account for two of the most significant events impacting Olympic security in recent history – Eric Rudolph's bombing of the Atlanta Games in 1996 and the 11 September 2001 attacks. In response, the following discussion will examine the role of surveillance technology across two distinct periods: 1976–2001 and 2001–present.[4] In making this distinction, however, it is important to recognise that many post-9/11 surveillance strategies are continuations and escalations of existing approaches (see, for example, Ball and Webster, 2004; Fussey, 2007b). Perhaps the most striking development since 9/11, however, is the scale intensity and scope of these practices, particularly in relation to the Olympics. Thus, many security motifs have been continually reasserted, yet their intensity has heightened. This is mirrored in the spiralling costs of securing such events. For example, Athens' (2004) security operation cost US$1.5bn, five-times that of Sydney (Coaffee and Fussey, 2010). Unravelling this figure reveals further points of interest. Whilst the number of athletes at Sydney and Athens was fairly similar (10,651 and 10,500 respectively), the costs of securing them contrasted sharply. According to one report, athletes attending Sydney received US$16,062-worth of security, whilst for those competing in Athens it had risen to US$142,857 (Hinds and Vlachou, 2007). Equally striking is the cost of security for each visitor, shifting from $34 to $283 per respective visitor to Sydney and Athens.[5] As the discussion below argues, these spiralling costs can be connected to growing contemporary 'protectionist reflexes' (Beck, 1999: 153) that attempt to create hermetically secure environments that place a central emphasis on technological modes of security such as surveillance. Key here is the way the distanciated nature of surveillance may provide a mechanism for Olympic organisers to negotiate the IOC's stated aim of accenting the Games as an athletic event and not an exercise in security, whilst responding to myriad threats and fears. In unpacking this process, the remainder of the chapter will first examine the anatomy of mega event surveillance strategies before 2001, before turning to an analysis of recent practices and innovations. The chapter will then draw out and review key points of analysis that arise from these strategies and processes.

Surveillance, sport and the Olympics before 2001

> it was agreed that the best way to deter suspected trouble-makers was ... [one] that would leave no doubt in their minds they were under continual close surveillance.
>
> (Official Report of the XXI Olympiad, Montreal, 1976: 559)

Sharply contrasting Munich's 'low key' approach to security (reflecting contemporary German sensitivities over conspicuous public displays of social control), little expense was spared on securing the 1976 Olympiad in Montreal. The global reaction to the kidnap and murder of 11 Israeli athletes and coaches by the Palestinian Black September Organisation meant that protection from terrorism had become a key concern for Montreal's organising committee (COJO). Although official noises made reference to the 'discreet efficiency' of Montreal's strategy (COJO, 1976), enormous emphasis was placed on three specific strands of security: a heavy accent on preventative measures, a strong and visible presence of security forces, and particular emphasis on enhanced and integrated communication and decision making measures (which had been a major failing at Munich and, later, at Atlanta). Surveillance plays a key role in augmenting and facilitating all three of these approaches. Specific measures included a seven week transport security corridor enabling athletes and dignitaries to be isolated on their journeys between sites, accommodation and transportation hubs, enhanced accreditation requirements for site workers (ibid.), acoustic surveillance measures installed in the Olympic village (Clarke, 1976, cited in Sanan, 1996a) and, crucially for this discussion, perhaps the first widespread and systematic deployment of CCTV to feature at an Olympics (COJO, 1976). Costing US$100m (equivalent to around US$380m today), the measures deployed at Montreal became staple features of subsequent Olympic security strategies and marked the increasing prominence of electronic surveillance.

A number of analytical points relevant today arise from the approach to security at Montreal. First is the issue of the continual mutability of threat. Many of the above measures are rooted in the fallout from Munich. Notwithstanding a recently lapsed threat from the ethno-nationalist Front de Libération du Québec (FLQ), questions may be raised concerning the efficacy and appropriateness of planning around retrospective events (particularly given that terrorists have far greater tactical preference for conventional bombing over hostage-taking across groups according to the Global Terrorism Database). Furthermore, despite the unprecedented expense and contemporary sophistication of its security operation, both the Olympic village and VIP protection were breached at Montreal. With the help of compatriots, an athlete succeeded in smuggling in and sheltering a friend in the Olympic Village for several days despite its comprehensive security cordon. In another incident, a foreign journalist succeeded in overcoming the VIP security ring, approached Queen Elizabeth and handed her a piece of paper as she awaited transportation from an Olympic related ceremony (COJO, 1976). Together, these events demonstrate the difficulties in achieving 'total protection'.

The baton of technological innovation was transferred to the next Olympiad, the XIII Winter Games at Lake Placid, New York State, in 1980. Here, organisers and security practitioners sought to improve on many of the strategies applied at Montreal and, in doing so, deployed the most advanced technological measures ever used at an Olympics. Taking advantage of the particular geography that allows many Winter Games to be physically separated from their surrounding environment more easily, many of the innovations involved the surveillance and strengthening of perimeters. In doing so, 12 ft high touch-sensitive fencing, voice analysers, 'bio-sensor' dogs, ground radar, night vision and CCTV were installed (Lake Placid Organizing Committee, 1980). Together, these represented innovations on surveillance and security practices used to secure military sites and airports, and subsequently became strategies emulated at later Olympics. Such was the securitisation of this environment, the legacy of the Olympic village was its conversion into a correctional facility (ibid.). These elements have since become templates that have informed and comprised the core of all subsequent Olympic security plans. Another innovation of the 1980 Winter Games' strategy was the use of private security. Although deployed at previous events (such as the 1964 Summer Games in Tokyo), private security agents were deployed extensively at Lake Placid and became a prominent feature at the United States' three subsequent Olympiads (Los Angeles, Atlanta and Salt Lake City) and elsewhere (notably at Seoul in 1988). This use of non-state-sanctioned actors connects with a number of operational issues relating to standards and integration in addition to ethical debates concerning accountability, legitimacy and transparency which are discussed below (pp. 105–111).

At the same time, CCTV was becoming a more familiar spectator at many of Britain's football stadia (following a feasibility study of camera surveillance at grounds in the West Midlands during the 1970s (Hancox and Morgan, 1975)). Its ubiquity was then assured as the Football Trust financed its installation at all British football clubs throughout the 1980s and 1990s, as well as providing funding for police forces to purchase mobile camera equipment (Armstrong and Giulianotti, 1998). Whilst the Football Trust articulated CCTV's utility in maximising spectator safety (although, see below, pp. 98–99), its main driver was the prevention of fighting or other disorder at the grounds – a fact evinced by the acceleration and escalation of CCTV deployment following the Heysel Stadium disaster of 1985.[6] Such was the extent of football-related CCTV surveillance by

1993, that an official Home Office Report on policing football hooliganism stressed that 'football supporters are probably more accustomed to being subjected to camera surveillance than most other groups in society' (Home Office, 1993, cited in Marsh *et al.*, 1996). Football supporters would therefore need to consent to being filmed in order to watch the match. In effect, surveillance and the sporting spectacle, observing and being observed, had become interconnected.

Other developments in the use of camera surveillance occurred during the mid 1990s, around the time that England last hosted what could be argued to be a mega sporting event – the 'Euro 96' football championships. A month before the championships were due to commence, on 5 May, a number of minor disturbances broke out in Newcastle upon Tyne, one of the host cities, as the city's football team had failed to win the Premier League title having surrendered a once substantial lead to Manchester United, the eventual winners. In response to the disorder, the police launched 'Operation Harvest' that drew on footage from 16 city-centre CCTV cameras and proceeded to conduct dawn raids on 30 homes culminating in the arrest of 19 individuals. At the time, this was claimed to have been the biggest policing operation ever to have utilised CCTV technology (Bennetto, 1996). Of interest here is the way entirely separate CCTV systems were becoming co-opted into more connected meta networks. Another component of these assemblages also became apparent during the actual Euro 96 championships – the public – as the latent utility of potential public participation was brought into play. Here, a technological variation on the long tradition of displaying suspects' pictures in public saw the media publication of CCTV stills of those suspected of being involved in public disorder. Another important component of the Euro 96 surveillance strategy was that of mobility. Mobile surveillance vehicles, so-called 'hoolivans', were extensively deployed throughout the championships (Marsh *et al.*, 1996) and have now become a central component of football policing strategies internationally, as evinced at the 2006 FIFA World Cup in Germany (Klausener, 2008) and 2008 European Championships in Austria and Switzerland (Hagemann, 2008).

Together, these three components – assemblage, mobility and the social environment – demonstrate that not only is technological surveillance a multifaceted phenomenon, but by itself it is limited (see, for example, McCahill, 2002) and dependent on its operational context. One event that both demonstrates some of the limitations of this technology at football grounds and underscores the importance of its social environment is the use of CCTV during the Hillsborough tragedy of April 1989. Cameras monitoring the congestion outside the ground led to decision to open gates at the Leppings Lane end of the stadium. Without diverting the flow of fans away from the overcrowded central pens towards the less crowded enclosures to the side of the goal, a fatal crush ensued (Home Office, 1989) which observers of the ground's CCTV feeds initially misinterpreted as hooliganism (Armstrong, 1998). In the immediate aftermath, Chief Superintendant Duckenfield, in overall command of the policing operation, is reported to have drawn on CCTV in an imprecise attempt to attribute culpability to disorderly supporters as he pointed to one of the control room's CCTV monitors and wrongly claimed, 'That's the gate that's been forced: there's been an inrush' (Home Office, 1989: 17). The role of CCTV in these tragic events demonstrates that deploying surveillance systems on their own not only limits their effectiveness but, also, can generate unforeseen circumstances. Additionally, the misinterpretation of CCTV images of the Hillsborough crowd suggests that these forms of surveillance technologies may be considered apart from other forms of technological information gathering, such as biometric surveillance or 'smart' ID cards. In these latter modes, through their very existence the technology itself asserts a function by capturing data. With CCTV, such 'capture' is not necessarily assured and enables operators to apply an intuitive gloss to its data. Thus, for CCTV, it is not necessarily the technology that does the work, but the contingent responses of surrounding actors (compete with their human frailties).

Overall, Olympic security between 1976–2001 can be seen as a period of both reaction and evolution. Practices and processes associated with securing these events became more standardised and internationalised during this time. At the same time, mainstream surveillance strategies became more prevalent and central, mirroring their deployment in more routine urban environments.

Surveillance and major sporting event security after 2001

Since 2001, Olympic and major sporting event security strategies have continued and built upon many of the themes present in earlier events. These include the aforementioned strands of target hardening, ambitions for achieving the 'total security' of designated spaces and the emphasis on technological surveillance in achieving this. What has changed since the turn of the millennium generally, and 9/11 specifically, has been the scale, technological innovation and centrality of surveillance strategies – in sum, an *intensification* (Ball and Webster, 2004; Fussey, 2007b). These features are visible in strategies designed to secure a series of key sporting events during this period as evinced by the following case studies discussing the 2001 Superbowl, the 2004 Olympic Games in Athens and the recent Beijing Olympics.

Superbowl XXXV, 2001

One key event in the evolution of surveillance technology occurred in January 2001 during the XXXV Superbowl at the Raymond James Stadium in Tampa. Here, facial recognition CCTV was deployed to monitor the crowd by taking pictures of every ticketholder as they passed through the turnstiles. Algorithms were then applied to the received images that measured facial characteristics such as the distance between the eyes, nose and lips, and the width of the nose. This information was then compared against a(n undisclosed) database in an attempt to efficiently identify those matching a similar facial profile belonging to offenders. Of the 100,000 attendees, the software flagged 19 individuals and was seen as a success by both the authorities (*Los Angeles Times*, 2001) and industry representatives (Lyon, 2003a). In response, the Florida police rolled the technology out into the public spaces of another part of the city, the Ybor City district.

However, the difficulties of operationalising this technology quickly tempered this enthusiasm. The chief difficulty here is, as Lyon (2003a) correctly notes, there is a world of difference between the demands on (and, hence, capabilities of) such technologies to *confirm* an identity (for example, at an airport check-in) and using them to *establish* an identity (for example, picking out a face in the crowd). Indeed, of the 19 individuals flagged, some were false alarms and none had committed anything more serious than ticket touting (ACLU, 2003). The subsequent experiences in less controlled public spaces also encountered difficulties. Following a public records request, research undertaken by the American Civil Liberties Union (2003) (enabled by Florida's open records law) revealed that Ybor City's FRCCTV had never identified anyone on its database and, further, was abandoned within three months of its deployment. Such experiences mirror those of the former US Immigration and Naturalization service who trialled and similarly abandoned the technology at the US–Mexican border. Such experiences ultimately led to FRCCTV being rejected at the first post-9/11 Olympics, in 2002 at Salt Lake City, Utah (although a version of this technology was deployed in non-public locations, such as vaults, to verify – not establish – identities).

While much of the opposition to FRCCTV has centred around privacy issues, another pressing debate concerns its operational limitations. One of the central difficulties concerns the sensitivity of such systems. To generate higher numbers of matches between what the cameras observes and its archived images, the sensitivity of the system needs to be increased. However, this is cantilevered by a commensurate increase in inaccurate 'identifications' – or 'false positives' (FRVT, 2006). Such incidences of 'false positives' sizeably outstrip the number of 'true positives' (ibid.) and, as such, subvert the original automated filtering intention and precipitate a more resource-intensive strategy. Compounding this is the impact of increased levels of scrutiny upon innocent people, which may ultimately undermine the legitimacy of enforcement agents.

A converse problem to false positives is that of false negatives, where the systems fail to identify those it is required to. This is due to the difficulties of operating in unpredictable and messy real world environments. For example, even amongst the best performing systems, variations in lighting can cause a severe deterioration in FRCCTV's capabilities. FRCCTV effectiveness has also been shown to deteriorate over time (because automated systems cannot necessarily compensate for ageing effectively) (FRVT, 2000). The technology is also hampered by the angle of the subject in relation to the camera – for example, one study notes a 15 degree angle led to a drastic drop in effectiveness, whilst facing the camera obliquely at more than 45 degrees rendered the cameras ineffective (FRVT, 2000). Moreover, the more a system is deployed, the less effective it becomes. For example, when databases have more images to compare, it reduces the capacity to distinguish between them as effectively (Introna and Wood, 2004).

Incorrect identification is not the only difficulty with such technology. As Introna and Wood (2004) note, whilst this technology is seen as neutral and unproblematic, certain biases are built into the algorithms governing their function. Amongst these is the tendency for FRCCTV to be more adept at identifying individuals from particular demographic groups depending on age, race and gender. In particular, older people are more easily detected than younger people, males more than females and those belonging to East Asian and African demographics are more easily identified than white Caucasians (ibid.).

Overall, in contrast to its proponents' claims, FRCCTV has yet to be proven to work in the ways its proponents have claimed.[7] Rather than this representing a fairly benign problem, the above discussion argues that security strategies could be undermined and the risks of victimisation potentially increased by their technological failure (an issue also germane to the technology-saturated strategy aimed at protecting the 2004 Athens Olympic Games, see below). In addition, such approaches consume finite resources that could be spent more effectively on low-tech and 'human' interventions that can interpret complex data, link security networks together and are flexible enough to adapt to continually changing environments.

Athens, 2004

Despite being the smallest country to host Olympics since Finland in 1952, the Greek Olympics of 2004 set out the most expensive, elaborate and extensive security programme ever deployed at the Games. The first post-9/11 summer Olympiad provides an exemplar (and possibly the apotheosis) of the 'total security' paradigm. Quintupling Sydney's security costs, $1.5bn was expended on the Athenian security project (Coaffee and Fussey, 2010). In part, some of this unprecedented figure may be accounted for by the limited extant security infrastructure at the time of bidding (when compared to, say, London or Beijing), yet the more direct drivers of this cost can be identified as both post-9/11 perceptions of vulnerability and demands for protection in addition to the heavy commitment to technological surveillance. Initially, this is borne out by the unprecedented levels of international military cooperation and the fact that, as Samatas (2007) notes, one-third of this figure funded electronic surveillance measures.[8] Together, this represents a heavy commitment to preventative measures – a theme that has been at the forefront of Olympic security planning since Montreal.

Dubbed by some as the 'Olympic superpanopticon' (Samatas, 2007), the Athenian surveillance system was monumental in scale. The flagship development was the 'C4I' ('Command', 'Control', 'Coordination', 'Communications' and 'Integration', ATHOC, 2004) programme comprising a network of 29 subsystems implemented by the US Science Applications International Corporation (SAIC) who had previously worked on the 2002 Winter Olympics. Much of the central focus of this system was on communication between the 135 command centres and disparate range of security operatives policing the Games. Of the explicitly surveillance-related provisions, the 35 Olympic sites and attendant critical national infrastructure were observed by an assemblage of thousands of CCTV cameras (estimates range from 1,470 in the official ATHOC (2005) report to almost 13,000 according to Coaffee and Johnston, 2007), at least one every 50 metres at Olympic installations; vehicle-tracking devices and dedicated CCTV subsystems monitoring roads, motion detectors, underwater sensors guarding Piraeus harbour; deployment of NATO AWACS aircraft; and a leased airship equipped with five cameras that could wirelessly transmit footage to major command centres (Hinds and Vlachou, 2007; Samatas, 2007). This was complemented by more traditional low-tech information and intelligence-gathering strategies. From 2000, a seven country Olympic Advisory Group facilitated international intelligence sharing (with the UK adopting a coordinating role) and the opening of an Olympic Intelligence Centre in the months leading up to the Games (Hinds and Vlachou, 2007).

Despite being the most technologically sophisticated security system ever deployed at an Olympics, and being supplied by a company (SAIC) that had trialled and operated systems at two previous Games (Sydney and Salt Lake City), the technological component of the Athens security operation was unequivocally problematic. These failures were not only articulated by the Greek government when seeking judicial arbitration for SAIC's failure to complete work nearly four years after the 2004 Games ended (Samatas, 2007), but also by the provider's admission of their own 'poor performance' (Bartlett and Steele, 2007).[9]

Athens' Olympic security system experienced difficulties from the start. Owing to a combination of the convoluted open tender process and delays from local construction and telecommunications companies, the contract was awarded eight months after work was due to commence.[10] Following this, the physical command centres were largely completed on time but, fatally, the software that managed the communications and governed the integration of security failed. Part of the issue related to capacity – despite 800 operators registered to use the C4I system, technical reports indicated that if more than 80 logged on at the same time, the system crashed, thus rendering it 'operationally useless' (cited in Samatas, 2007: 228). What is perhaps more significant is the human and social response to this technological failure. For example, officials supplemented the network by adding more guards and the Greek military, in (correctly) anticipating that the C4I system would not be fully operational in time, deployed its own communication systems (ibid.).

An important operational point related to the relationship of human innovation and technological networks is raised here. Technological failures may not just represent the unfortunate denial of an additional service but, owing to the social environment these strategies are deployed within, may incur additional risks.[11] For example, deploying an alternative communication network to one component of a security coalition amplifies their isolation and decreases the prospects of successful communication and integration.

Beijing, 2008

> Without security guarantees there cannot be a successful Olympic Games, and without security guarantees the national image will be lost.
>
> (Hu Jintao, cited in Macartney, 2008)

The standardisation of Olympic security can also be observed in the less likely setting of Beijing. Hinds and Vlachou (2007) have argued, for example, that much of the strategising for the 2008 Games was informed by planning for the 2006 Winter Olympics in Turin, the summer Games in Athens, 2004, and Melbourne's hosting of the 2006 Commonwealth Games. Whilst certain features and motifs may have been transferred, it is also important to note that a number of features specific to China played a substantive role in shaping and propelling the Olympic security strategy.

The IOC's decision to award the XXIX Olympiad to Beijing in July 2001 stimulated a monumental programme of Olympic related redevelopment. Although objective statistics on the overall expenditure on the Beijing Games are difficult to verify, estimates suggest that $40bn was spent upgrading the city's infrastructure, with $23bn of that figure ring fenced for specific Olympic development (Evans, 2007). What perhaps separates Beijing's redevelopment experience from the preceding (and probably succeeding) Olympiads is the particular confluence of global and national processes at that specific time. In one respect, Beijing's strategy was facilitated by the state's immense power to mobilise security (as experienced by the totalitarian Moscow and reforming Seoul Olympiads). At the same time, 'Dengist' notions of 'socialism with Chinese characters' (Cook, 2007) – or a state oriented yet liberalised economy receptive to globalisation – enabled the infrastructure and machinery of security to be imported whilst requiring specific brand images of the city to be exported.[12]

Accompanying this redevelopment was a sizeable commitment to security that accented technological and surveillance based approaches. After being awarded the Games, the Chinese authorities accelerated their acquisition of hi-tech security apparatus. This enabled the deployment of technologically advanced surveillance measures during the Games, including the use of Radio Frequency Identification (RFID) tags in tickets to some events (such as the opening ceremonies) to enable their holders' movements to be monitored. Despite these headline catching technologies, however, the principal emphasis has been on developing CCTV networks. This is evinced by the 2001 launch of the 'Grand Beijing Safeguard Sphere' (running up to the start of the Games in 2008) predominantly focusing on the construction and integration of a city wide CCTV system that some sources (Security Products, 2007) claim cost over $6bn. This and related initiatives has led to estimates that Beijing now hosts over 300,000 public CCTV cameras (see, for example, Magnier, 2008). Also notable is the was China's recent trend towards hosting international mega events has driven the deployment of surveillance cameras across a number of other cities including Shanghai (relating to 'Expo 2010') and Guangzhou (the 2010 Asian

Games). These developments have further catalysed a nationwide 'Safe Cities' programme to establish surveillance cameras in 600 cities (Bradsher, 2007). Overall such developments, in all likelihood, have probably allowed China to claim Britain's dubious accolade of the planet's most intensely observed nation.

In addition to this burgeoning scale, these new installations arguably exert a much greater scope than the systems deployed to observe Britain's citizens. In contrast to the varied hardware quality and often ad hoc (sometimes partially atrophied) networks that constitute many UK systems (see, for example, McCahill, 2002; Fussey, 2007a), the simultaneous deployment of entire surveillance infrastructures across China's urban spaces brings a much higher degree of cohesion and, hence, capacity. In areas of exceptionally high population density, such as the 22,394 people per square kilometre in Beijing (Peng and Yu, 2008), observational capacity becomes additionally potent. Of further note here is the pernicious way that cameras installed for one purpose may be applied for more malign ends ('function creep'), as seen by the use of Beijing's Panasonic-supplied 'traffic cameras' to track down participants of the 1989 Tiananmen Square protests.[13]

Criticism of Beijing's surveillance strategies has largely concentrated on ethical considerations. In addition to the aforementioned Orwellian imagery evoked when a one party regime heavily invests in technological surveillance, criticism of Beijing's strategy has focused on more traditional concerns such as the restriction on protests, the 'overpolicing' of certain populations and the displacement of communities. Regarding the latter, for example, Achaya (2005, cited in Cook, 2007) argues that over 300,000 Beijing residents were evicted due to the Games. Whilst these issues are undeniably serious, it is important to point out that each of these criticisms also applies to the security practices adopted at many other Games across the globe. Restriction on protest in or near Olympic venues, for example, constitutes Rule 51, section 3 of the IOC's Olympic Charter that host cities are compelled to adhere to (and therefore may override or contradict enshrined domestic rights). The overpolicing of subpopulations has also been controversial in many societies and, in relation to the Olympics, was particularly divisive at Sydney, 2000 (see Lenskyj, 2002). Additionally, it is important to note that Western companies, particularly US technology giants, played a key role in providing much of the technological surveillance apparatus to China. According to the *New York Times* (Bradsher, 2007), these include General Electric's 'Visiowave' system that allows security officers simultaneous control of thousands of cameras, a Honeywell system to automatically analyse feeds from cameras deployed at Olympic sites and a similar IBM system to analyse and catalogue behaviour. What is controversial here is that such transactions can be argued to contravene the spirit (if not the letter) of key parts of the Foreign Relations Authorization Act passed by US Congress in 1990 (colloquially known as the 'Tiananmen Sanctions'). In response to the clampdown and executions connected to the Tiananmen Square protests, this Act prohibited sale of 'crime control or detection instruments or equipment' to China.

Reflecting on surveillance and the Olympic spectacle

Overall, this discussion has illustrated that, whilst technological surveillance has been an important feature of major-event security since Munich, it has gathered pace in both extent and sophistication during the post-9/11 era. This trend looks likely to continue in the run-up to the 2012 Games in London. Indeed, given the UK's long history of policing urban terrorism and spearheading the deployment of technological forms of crime control, there are arguably more expectations on security for 2012 than for countries hosting other Games. Moreover, the London Olympic Delivery Authority's (ODA) tenders for potential private sector 2012 security suppliers encourage applications that 'create an integrated security environment that is effective, discreet and proportionate' (ODA, 2007). This echoing of the IOC's longstanding (and often abandoned) aim of projecting the Games as an athletic event and not an exercise in security is significant as 'discretion' and 'proportionality' have normally translated into distanciated forms of control, particularly in the form of technological surveillance. Overall, what is therefore certain is that, for 2012, the importance afforded to technological surveillance will remain. Having discussed and critiqued the deployment and role of technological surveillance in securing major sporting events, this chapter now considers some of the overarching operational and critical issues associated with such securing sport through surveillance.

One of the key lessons emerging from Britain's long experience with camera surveillance is that technology does not provide a panacea for the various threats to security. Instead, its efficacy is governed by its operational and social context. These contexts provide a backdrop to the following thematic discussions.

Shifting landscapes of security and insecurity

Hosting the Olympics is a once in a generation undertaking. As such, for the host city, its attendant security provisions are likely to be exceptional in scale, composition and scope. To effectively negotiate this unprecedented challenge, the evidence suggests that particular security motifs have become standardised and transferred across events and nationalities. Whilst the mechanisms for this transfer remain complex (involving mediums such as networks of security professionals (see Klausner, 2008) or the IOC taking an active role as an 'institutional memory'), consolidated, generalised and transferable paradigms have become routinely applied across diverse settings. A number of analytical points are raised by such practices.

In addition to the variegated threat, standardised security strategies may not function effectively in disparate operational environments. This problem was documented at the 2002 Winter Olympics in Salt Lake City, where sophisticated communications technology supplied by federal agencies failed to work in Utah's mountainous terrain (Decker *et al.*, 2005). Moreover, when technologies do work, they may operate differently in diverse contexts. For example, proponents of CCTV have consistently cited the utility of surveillance cameras in stopping a criminal act before its commission via its deterrence effects – a feature that has some evidential basis with regard to preventing property and vehicle crime (see, for example, Tilley, 1998), if not more serious interpersonal crimes. With reference to terrorism, this utility often shifts to the post-event setting. Although this relationship has been explored elsewhere in greater depth (Fussey, 2007b) than can be addressed here, a number of examples can be cited to illustrate the point. Indeed, one of the first uses of public space CCTV to counter terrorism in mainland Britain was in this post-event capacity, when it was employed to assist with the identification of two English-born members of PIRA who planted a bomb outside Harrods in Knightsbridge, London, in March 1993. More recently, iconic surveillance camera footage has been published following 9/11 and the 7 July and (failed) 21 July 2005 London attacks.[14] Where CCTV has been successful in preventing terrorist attacks, this has often followed information gathered from previous attacks in the same campaign, as evinced by the capture of David Copeland in 1999 following his detonation of three nail bombs in London during the same year.

Supplying security?

Another area of discussion concerns whether these surveillance technologies provide the security that they promise. Indeed, considerable debate has raged over the effectiveness of the most prominent surveillance technology – CCTV. For advocates, one of the major strengths of CCTV lies in its versatility and potential for multiple applications, primarily in terms of its deterrent and detection capabilities. However, critics have pointed out that the potential for surveillance to work does not necessarily translate into its actual efficacy (see Tilley, 1998, for a detailed discussion on this point). Additionally, even though CCTV may be applicable in many contexts, its performance is unlikely to be commensurate in all tasks. Like many crime control strategies, claims and counter-claims of surveillance strategies are commonplace. With regard to CCTV, many evaluations of its effectiveness conflict on the same point. As Tilley (1998: 140) states, 'mixed findings are the norm'. Such conflict is clearly articulated in one Home Office-funded meta-analysis of 22 studies of CCTV effectiveness.[15] Here, Welsh and Farrington (2002) contend that, for each study claiming CCTV effectiveness, another study exists pointing out an inconclusive or negative effect. Despite such uncertainty, there is some evidence to suggest that CCTV can be effective, but only in certain circumstances. Specifically, there is broad consensus that CCTV is most effective in tackling vehicle crime (an enduring Home Office priority) whilst having little (if any) real effect on violent crimes (a greater cause of fear of crime) (see, for example, Tilley, 1993; Welsh and Farrington, 2002). Other studies have even identified an *increase* in violent offences in some locations following the deployment of CCTV (Gill and Spriggs, 2005).[16]

The limitations of surveillance strategies are more pronounced in relation to second-generation technologies, as noted above in the case of FRCCTV.

Nevertheless, technological surveillance measures aimed at pre-empting terrorist attacks have become a central feature of counter terrorism strategies since 9/11 (see, for example, Lyon, 2003a). Indeed, as Gates (2006) notes, in the US every major piece of post-9/11 federal security legislation has included biometrics provisions. Such developments have been strikingly accompanied by manufacturers' claims that the 9/11 attacks would not have occurred had particular variants of biometric surveillance (such as facial recognition) been in place (ibid.). The fact that in that particular case it was not establishing identity that was a problem but, rather, determining intent, renders such claims as both hubristic and incorrect. The same problem stubbornly asserts itself in other contexts. Indeed, were FRCCTV installed and, hypothetically, working effectively at Luton station on 7 July 2005, it may have identified Mohamed Siddique Khan, Shezad Tanweer, Jermaine Lindsay and Hasib Hussein as who they were, but not invested their passing through the station with any meaning or understanding of intent.

One of the important dynamics affecting the deployment of surveillance and Olympic security is that of displacement (see also Silke, this volume). Displacement refers to innovative aspects of transgressive behaviour that attempt to circumvent defences rather than overcome them. Criminological research has indicated how this can occur in numerous ways including, amongst others, the decision to attack a different location (spatial displacement), at different times (temporal displacement) or adopt different techniques (tactical displacement). With regard to the Olympics, attacks on and during the Games are rare, whilst those related to the Games, yet which are temporally and spatially distinct, are more common.[17]

New technologies also affect human behaviour, causing individuals to respond and adapt in new and often unpredictable ways. One fairly benign example of this process of technological displacement is the development and deployment of 'mosquito alarms' to discourage adolescents from congregating in specific areas. Based on the physiological condition of presbycusis – where the aging process limits the range of sound frequencies audible to adults – mosquito alarms emit a high frequency sound only perceptible to individuals younger than their mid twenties. Conversely, this has also provided schoolchildren with mobile phone ringtones that can operate beyond the aural range of many of their teachers. Although a flippant example, the underlying point is more serious: technological innovation stimulates human adaptability. One illustrative terrorism-related example is PIRA's dialectical responses to the British government's technology-informed attempts to curtail remote bomb detonations. As Hoffman (1994) notes, to counter PIRA's detonation of explosives via radio-control, the Ministry of Defence (MoD) developed methods of jamming the signals. PIRA circumvented this by developing a strategy comprising complex arrangements of switches that, in turn, was annulled by MoD-developed scanners that could identify and block signals in the short space of time they were being transmitted. PIRA eventually overcame this obstacle via the use of photographic flash 'slave units' which can be detonated remotely from hundreds of metres by operating conventional camera flashes (ibid.) – a method that has yet to be comprehensively countered.[18] Such innovations represent both the dynamic relationship between technology and terrorism and serve to illustrate how the utopian claims of technological solutions to terrorism need to be treated with caution. In sum, technology does not operate within a vacuum. Instead, its operation and efficacy are determined in relation to the human and social contexts it functions within.

Scale and reach

For their hosts, the Olympics normally entail the largest security they are likely to face. Beyond the need to account for a range of myriad conventional and unconventional risks, Olympic sized operations have normally required temporary coalitions of numerous agencies to work together, often for the first time (in the case of Atlanta's security programme, this involved the input of over 40 agencies (ACOG, 1997)). For the deployment of surveillance technologies within these large-scale overarching strategies, two important issues arise – those of coordination and overload. Both of these further accent the importance of technology's social environment in governing its efficacy.

Regarding the former, issues of coordination and effective integration have been key determinants of the successes and failures of Olympic security operations. As has now been well documented (see, for example, Sanan, 1996a, for a comprehensive discussion), the attempt to resolve the Munich hostage situation by force was compromised by poor communication (including a misrepresentation of the number of BSO terrorists) between the Bavarian and Federal police. Atlanta's decentralised approach to Olympic security may have also contributed to the impact of Eric Rudolph's crude pipe bomb. As has now been well-documented (see, for example, Buntin, 2000), the Atlanta Police Department received a bomb threat 30 minutes before the explosion yet did not effectively pass the information to the specific Olympic bomb management unit for 20 minutes. The bomb was eventually discovered independently and exploded during the process of evacuation. More recently, other problems in coordinating Olympic strategies occurred at Athens, 2004, as discussed above. Together, these examples illustrate that surveillance measures are limited if their observations are not communicated effectively. As such, the tension between scale and cohesion constitutes one of the most crucial elements of Olympic security.

The second issue related to the unprecedented scale of Olympic surveillance operations concerns data overload. As technological surveillance infrastructures expand, exponential amounts of data are generated. Notwithstanding differences in the dynamics of the attacks and shifts in investigation procedures, one example of how this may occur in relation to terrorism in London is to consider the comparative volumes of CCTV footage following David Copeland's nail-bombing campaign of 1999 and the 7 July 2005 attacks of the city's transport infrastructure. Following the Copeland case, the Metropolitan Police analysed around 26,000 hours of footage taken from public CCTV cameras over his three-week bombing campaign (Metropolitan Police, 1999). Following the 7 July bombings six years later, the Metropolitan Police Anti-Terrorism Branch's (SO13) request for all CCTV material from one single London borough, Hackney, for the hours between midnight and 1 pm yielded 22,015 hours of footage, equivalent to 56,000 CDs of data (Wells, 2007). There are 32 London boroughs.

This issue of overload is evident in other types of surveillance measures, particularly second generation technologies designed to sort and filter data. This is because automated forms of surveillance still generate large volumes of data. In London, one key example is the way in which probably the most straightforward form of this technology – ANPR – has increased police workloads. Such has been the growth in data volumes from these cameras, it has led the Police Authority to request an increase in the Police Precept (the quota of council tax revenues apportioned to the police) to cover the large human resource shortfalls caused by the labour intensive review of this aggrandised data (Sherman and Ahmed, 2008). This mushrooming of available information places additional pressures on the operational environments of those utilising surveillance technologies. Together, such factors indicate the need to balance increasingly scarce and finite resources towards enhancing the capacity of human agents to analyse and interpret available information over investment in technological mechanisms to generate additional data.

Together, such examples illustrate that technology does not operate within a vacuum. Instead, its operation and efficacy is determined in relation to the human contexts it functions within. Human agents design the algorithmic focus of surveillance technologies, manage their operation and interpret the outputs, even those of 'second generation' machines. As such, the operation of newly inaugurated technologies is affected by unpredictable human contingencies. Simultaneously, new technologies also impact and, to some extent, shape their social environment by altering the way in which people work and operate.

Surveillance and ethics

Potential limitations in the effectiveness of these strategies also suggests that the enduring 'liberty vs security' dichotomy that frames much surveillance related discussion does not adequately capture the complexity of the issue. Primarily, this is due to the implicit assumption that these measures operate in an unfettered and effective manner to either provide security or deny liberty. This chapter argues that such perspectives, whilst important, only confront some of the relevant issues and, to move the debate along, it is necessary to address other, less-acknowledged yet equally important, issues. In doing so, five ethical considerations that should constitute tangible concerns for security planners are identified here as legacy, legitimacy, locality, listings and legislation, and are discussed in turn.

Legacy is perhaps one of the most frequently articulated concepts in relation to 2012, and is commonly associated with the post-event use of the

Olympic site and attendant community regeneration schemes. Regarding security, current tenders for 2012 Olympic Park security providers are encouraging companies to supply 'security legacy', thus bequeathing substantial mechanisms and technologies of control to the post event site. Here, questions remain over the security priorities of a high profile international sporting event attended by millions of people and the degree of infrastructure that will remain to police a large urban parkland (the future incarnation of the Olympic site). Although, for other mega events, such as the 2006 FIFA World Cup, the security legacy consisted of sustained networks of professionals rather than physical control measures (largely due to the deployment of mobile CCTV) (Baasch, 2008; Hagemann, 2008), this post-event inheritance of security infrastructures is a common Olympic legacy.

This connects with the second ethical issue, that of *legitimacy*. One of the most valuable areas of agreement across much policy, practice and research is that policing agencies require legitimising via the consent of the policed (see, for example, Reiner, 2007). However, many Olympic cities have experienced the post-event retention of policing agencies and strategies that the public did not initially consent to. These include the legacy of private policing following the Tokyo (1964) and Seoul (1988) Olympiads (Lee, 2004), and the continuation of zero-tolerance-style exclusion laws after Sydney (2000) (Lenskyj, 2002). Together, these two ethical themes connect with critical notions of control 'creep' (see, for example, Marx, 1988). In London, as elsewhere, the legitimacy of policing institutions has been long recognised as an important issue and is one that has involved a number of hard lessons (see, for example, Scarman, 1981; Macpherson, 1999). Given that the 'community' is continually cited by the ODA as the main benefactor of the Games, this perhaps constitutes one of the most important socio-ethical issues.

Also connected is the issue of *locality*. Mega-sporting events internationalise the local community and, in doing so, create a security environment aimed at responding to exceptional needs. As Lenskyj (2002) notes in the case of Sydney and Samatas (2007) in respect to Athens, this may stimulate the decline of public control over the local environment. Indeed, during the recent 2008 UEFA European football championships, Klausner (2008) explains how residents of one Swiss city expressed their opposition to CCTV deployment in two separate referendums. On becoming a host venue for the championships, UEFA's requests for the cameras led the municipal authorities to install them in spite of this democratically articulated public opposition.

Connecting with Lyon's (2003a) argument that the most important social cost of surveillance is the 'sorting' of individuals into categories that could substantially affect their life chances without available mechanisms of redress, the '*listing*' or classification of individuals as a risk has been a corollary of mega-event security practices in some countries. In addition to debates over the accuracy of pre-emptive and actuarial processes (Feeley and Simon, 1994) that define risk, one key issue here is how individuals extricate themselves from the categories or lists to which they have been allocated, particularly when the categorisation may be incorrect and/or have a significant effect on life chances and opportunities. For example, Eick and Töpfer's (2008) study of security at the 2006 FIFA World Cup in Germany revealed police made 8,450 'contacts' with people they suspected *may become* involved in hooliganism during the tournament. These contacts often took the form of uniformed officers questioning individuals at their workplace or family environment.

The final area of ethical debate concerns *legislation*. As noted above, hosting mega sporting events may generate tensions with domestic legislative arrangements. This can occur in a number of ways. Primarily, the international instruments of the IOC's Olympic Charter, particularly those connected to protest and assembly, may contrast with domestic commitments. Alternatively, as demonstrated by the procurement issues surrounding the Beijing Games, the spirit of domestic laws connected to the sales of military and policing apparatus may be bypassed in favour of the huge commercial and reputational benefits to suppliers of Olympic security equipment.

Together, these manifest issues affecting previous mega sporting events demonstrate how ethical issues extend beyond narrow conceptions of privacy and liberty to include a range of potential social costs. For Olympic sized events to provide a genuinely beneficial legacy to local communities, these issues need to be defined broadly.

This chapter has charted the increasing centrality of surveillance strategies within Olympic security operations since Munich. In doing so, a number of key critical issues have emerged. Notably, crime prevention orthodoxies, increasingly translated into counter-terrorism contexts, have become central features of the progressively globalised security operations protecting mega events. Amongst these, technological forms of surveillance have become central, achieving particular prominence since 9/11. Moreover, the chapter has argued that a number of issues affecting their more routine applications continue to be relevant in the exceptional and unprecedented context of such operations. In particular, although technological forms of surveillance have gained primacy, it is their human and social environments that remain crucial, yet under-acknowledged, factors mediating their efficiency and effectiveness.

Notes

1 Although the Rodney King beating was filmed by a personal video camera rather than CCTV. In an interesting development, Pecora (2002) notes how, since the Rodney King episode, many LAPD officers now carry their own video cameras to protect themselves from complaints of brutality or subversion of proper procedure.
2 Although distinctions are not made in the NACRO report, this presumably relates to the extent of CCTV in public space, indiscriminately placing the wider population under surveillance. McCahill and Norris (2004) refer to all cameras sited in both public and private locations. Moreover, some practitioners feel that McCahill and Norris' figure is actually an underestimate (Parkins, 2007).
3 Considerable debate exists here. As this author and others have argued, whilst the totalitarian nature of social control in Brezhnev's Russia enabled the use of security apparatus that may be unacceptable in other societies, this insularity does not

112   *P. Fussey*

necessarily mean that 1980 Olympic security operation was entirely distinct from other events across time and place (Fussey *et al.*, forthcoming). US-made security apparatus such as metal detectors and X-ray scanners were deployed (as they were at Lake Placid) and Moscow's zero tolerance style policing approaches have also featured at subsequent Games, notably Sydney (Lenskyj, 2002) and Beijing (Peng and Yu, 2008), albeit with variations of scale.

4 The key event determining this distinction is clearly 9/11, although prior to this event, a number of innovations were also introduced into sporting event security during this year, such as the introduction of FRCCTV at the Superbowl in Tampa.

5 Applying fairly simple mathematics to this estimation suggests the probability that Hinds and Vlachou have applied an inconsistent calculation to the costs of both Games. However, the point regarding the wide disparities between the cost of securing both events remains valid.

6 However, in the post-Hillsborough inquiry, Lord Justice Taylor criticised many of these initial systems as being of inadequate quality and staffed by poorly trained operators (Home Office, 1990: 48).

7 Some manufacturers have gone so far as to claim 9/11 could have been prevented had their products been deployed (Gates, 2006). Given both the limitations of this technology and the fact that it was the hijackers' intentions that were invisible, not their identities, this claim appears difficult to justify.

8 Another notable component of the Athenian security programme was the large numbers of personnel deployed.

9 Bartlett and Steele (2007) argue that this resulted in SAIC losing $123 million on the project.

10 According to some accounts, this delay in issuing contracts affected security planning for the 2006 Winter Games in Turin. Mills (2006) argues that, instead of employing local Italian companies to provide the technology, as originally intended, delays meant that companies with prior experience of providing such systems needed to be drafted in so that deadlines could be met. This meant the awarding of contracts to US companies that have worked on previous Games. This also illustrates another mechanism sustaining the transference of security practices across Olympiads.

11 Just as the failures of some surveillance technologies, such as FRCCTV, drained finite resources by addressing labour-intensive false positives.

12 C.f. BOCOG's official Beijing 2008 slogan, 'One World, One Dream'.

13 Of course, function creep is not restricted to Chinese CCTV systems. Many examples can be drawn from surveillance practices in the UK. Taking the City of London's so-called 'Ring of Steel' as an example, cameras originally installed to tackle the threat from PIRA bombs are more routinely employed to tackle vehicle crime (Coaffee, 2004) and, in an example of the potential creep from 'civil' to 'criminal' applications, the Home Office has investigated the possibilities for co-opting the automatic number plate recognition (ANPR) cameras installed to administer London's 'Congestion Charge' for counter-terrorist purposes. Whilst these appear to be reasonably benign examples, such applications raise ethical questions concerning purpose and legitimacy.

14 Although CCTV was incidental on 7/7, as the bombers were identified via other (more traditional) means (House of Commons, 2006).

15 These 22 studies were selected on the grounds of their methodological robustness and included criteria such as the comparison of the experimental area with a control area, the use of 'before' and 'after' studies of sufficient scope, and the use of CCTV as the most significant crime control intervention (so it can be distinguished from other interventions deployed at the same time).

16 However, such increases should not be altogether surprising, given the comparative escalation of violent crime in some areas of the UK in recent years. There are also methodological reasons for the statistical increase of violent offences.

17 It could also be argued that Eric Rudolph's attack on the 1996 Atlanta Games was partially displaced. Rudolph attacked the Games due to his opposition to what he saw as the US government's pro-abortion stance (Toohey and Veal, 2007). However, the fact that he attacked Centennial Park, a quasi-public (and therefore less-regulated) space, in preference to the fortified Olympic site may indicate the influence of spatial displacement.

18 For a different example documenting the subversion of biometric scanners at airports, see Lyon (2003b).

References

ACLU (2003) *Q&A on Face-Recognition*, available from www.aclu.org/privacy/spying/14875res20030902.html (retrieved 1 February 2010).

Armstrong, G. (1998) *Hooligans: Knowing the Score*, Oxford: Berg.

Armstrong, G. and Giulianotti, R. (1998) 'From Another Angle: Police Surveillance and Football Supporters', in C. Norris, J. Moran, G. Armstrong (eds) *Surveillance, Closed Circuit Television and Social Control*, Aldershot: Ashgate, pp. 113–138.

ATHOC (2005) *Official Report of the XXVIII Olympiad: Homecoming of the Games*, Athens: ATHOC.

Atlanta Committee for the Olympic Games (ACOG) (1997) *The Official Report of the Centennial Olympic Games*, Atlanta: ACOG.

Baash, S. (2008) *FIFA Soccer World Cup 2006: Event-Driven Security Policies*, Surveillance and Security at Mega Sport Events: From Beijing 2008 to London 2012 Conference, Durham University, 25 April.

Ball, K. and Webster, F. (eds) (2004) *The Intensification of Surveillance: Crime, Terrorism and Warfare in the Information Era*, London: Pluto.

Bartlett, D. and Steele, J. (2007) *Washington's $8 Billion Shadow*, available from www.barlettandsteele.com/journalism/vf_washington8_3.php (retrieved 1 February 2010).

Beck, U. (1999) *World Risk Society*, London: Polity Press.

Bennetto, J. (1996) 'Euro 96 Violence Targeted by Police', *Independent*, 25 May, available from www.independent.co.uk/news/euro-96-violence-targeted-by-police-1349003.html (retrieved 1 February 2010).

Bradsher, K. (2007) 'China Finds American Allies for Security', *New York Times*, 28 December.

Buntin, J. (2000) *Security Preparations for the 1996 Centennial Olympic Games (B): Seeking a Structural Fix*, Kennedy School of Government Case Program, Harvard.

Carter, H. (2004) 'Inquiry as CCTV Captures Police Beating', *Guardian*, 21 February, available from www.guardian.co.uk/crime/article/0,2763,1152990,00.html (retrieved 1 February 2010).

Coaffee, J. (2004) 'Recasting the "Ring of Steel": Designing Out Terrorism in the City of London', in S. Graham (ed.) *Cities, War and Terrorism: Towards an Urban Geopolitics*, Oxford: Blackwell.

Coaffee, J. and Fussey, P. (2010) 'Security and the Threat of Terrorism', in J. Gold and M. Gold (eds) *Olympic Cities: City Agendas, Planning, and the World's Games, 1896 to 2012* (Second Edition), London: Routledge.

Coaffee, J. and Johnston, L. (2007) 'Accommodating the Spectacle', in J. Gold and M.M. Gold (eds) *Olympic Cities: Urban Planning, City Agendas and the World's Games, 1896 to the Present*, London: Routledge.

COJO (1976) *Official Report of the XXI Olympiad, Montreal 1976*, Ottawa: COJO.

Cook, I. (2007) 'Beijing, 2008', in J. Gold and M. Gold (eds) *Olympic Cities: City Agendas, Planning, and the World's Games, 1896 to 2012*, London: Routledge.

Davis, M. (1998) *Ecology of Fear: Los Angeles and the Imagination of Disaster*, New York: Metropolitan Books.

Decker, S., Greene, J., Webb, V., Rojeck, J., McDevitt., Bynum, T., Varano, S. and Manning, P. (2005) 'Safety and Security at Special Events: the Case of the Salt Lake City Olympic Games', *Security Journal*, 18(4): 65–75.

Ditton, J. (1999) 'CCTV and Public Space', paper given at British Criminological Society meeting, Loughborough University.

Eick, V. and Töpfer, E. (2008) 'The Human and Hardware of Policing Neoliberal Sport Events', paper presented at *Surveillance and Security at Mega Sport Events: From Beijing 2008 to London 2012*, Durham University, 25 April.

Evans, G. (2007) 'London 2012', in J. Gold and M. Gold (eds) *Olympic Cities: City Agendas, Planning, and the World's Games, 1896 to 2012*, London: Routledge.

Feeley, M. and Simon, J. (1994) 'Actuarial Justice: the Emerging New Criminal Law', in D. Nelkin (ed.) *Futures of Criminology*, London: Sage.

Foucault, M. (1977) *Discipline and Punish: the Birth of the Prison*, London: Penguin.

FRVT (2000) *Facial Recognition Vendor Test 2000*, available from www.frvt.org/FRVT2000/default.htm (retrieved 30 June 2009).

FRVT (2006) *Facial Recognition Vendor Test 2006*, available from www.frvt.org/FRVT2006 (retrieved 30 June 2009).

Fussey, P. (2007a) 'An Interrupted Transmission? Processes of CCTV Implementation and the Impact of Human Agency', *Surveillance and Society*, 4(3): 229–256.

Fussey, P. (2007b) 'Observing Potentiality in the Global City: Surveillance and Counter-terrorism in London', *International Criminal Justice Review*, 17(3): 171–192.

Fussey, P., Coaffee, J., Armstrong, G. and Hobbs, R. (in press) *Securing and Sustaining the Olympic City: Reconfiguring London for 2012 and Beyond*, Aldershot: Ashgate.

Fyfe, N.R. (2004) 'Zero Tolerance, Tolerance, Maximum Surveillance? Deviance, Difference and Crime Control in the Late-Modern City', in L. Lees (ed.) *The Emancipatory City? Paradoxes and Possibilities*, London: Sage.

Gates, K. (2006) 'Identifying the 9/11 "Faces of Terror": the Promise and Problem of Facial Recognition Technology', *Cultural Studies*, 20(4–5): 417–441.

Gill, M. and Spriggs, A. (2005) *Assessing the Impact of CCTV*, Home Office Research Study no. 292. Home Office: London, available from www.homeoffice.gov.uk/rds/pdfs05/hors292.pdf.

Graham, S. (1998) 'Towards a Fifth Utility? On the Extension and Normalisation of Public CCTV', in C. Norris, J. Moran and G. Armstrong (eds) *Surveillance, Closed Circuit Television and Social Control*, Aldershot: Ashgate.

Hagemann, A. (2008) 'From Stadium to "Fanzone": The Architecture of Control', paper presented at *Surveillance and Security at Mega Sport Events: From Beijing 2008 to London 2012*, Durham University, 25 April.

Hancox, P.D. and Morgan, J.B. (1975) 'The Use of CCTV for Police Control at Football Matches', *Police Research Bulletin*, 25: 41–44.

Hinds, A. and Vlachou, E. (2007) 'Fortress Olympics: Counting the Cost of Major Event Security', *Janes' Intelligence Review*, 19(5): 20–26.

Hobbs, D. (1988) *Doing the Business*, Oxford: Oxford University Press.

Hoffman, B. (1994) 'Responding to Terrorism Across the Technological Spectrum', *Terrorism and Political Violence*, 6(3): 366–390.

Horne, C (1996) 'The Case For: CCTV Should be Introduced', in *International Journal of Risk, Security and Crime Prevention*, 1(4): 317–326.
Home Office (1989) *The Hillsborough Stadium Disaster, 15 April 1989: Inquiry By The Rt Hon Lord Justice Taylor. Interim Report*, London: HMSO.
Home Office (1990) *The Hillsborough Stadium Disaster, 15 April 1989: Inquiry By The Rt Hon Lord Justice Taylor. Final Report*, London: HMSO.
House of Commons (2006) *Report of the Official Account of the Bombings in London on 7th July 2005*, London: HMSO, available from www.homeoffice.gov.uk/documents/7-july-report.pdf?view=Binary (retrieved 1 February 2010).
Introna, L.D. and Wood, D. (2004) 'Picturing Algorithmic Surveillance: the Politics of Facial Recognition Systems', *Surveillance and Society*, 2(2/3): 177–198.
Klauser, F. (2008) 'Exemplifications of Security Politics Through Mega Sport Events', paper presented at *Surveillance and Security at Mega Sport Events: From Beijing 2008 to London 2012*, Durham University, 25 April.
Lake Placid Olympic Organising Committee (1980) *Final Report of the XIII Winter Olympic Games Lake Placid, 1980*, New York: LPOOC.
Lee, C. (2004) 'Accounting for Rapid Growth of Private Policing in South Korea', *Journal of Criminal Justice*, 32: 113–122.
Lenskyj, H. (2002) *The Best Olympics Ever? Social Impacts of Sydney 2000*, Albany: State University of New York Press.
Los Angeles Times (2001) 'Super Day for Big Brother', *Los Angeles Times*, 2 February.
Lyon, D. (2003a) *Surveillance after September 11th*, London: Blackwell.
Lyon, D. (2003b) 'Surveillance as Social Sorting: Computer Codes and Mobile Bodies', in D. Lyon (ed.) *Surveillance as Social Sorting: Privacy, Risk and Digital Discrimination*, London: Routledge.
Macartney, J. (2008) 'Stop Blocking the Internet, Olympics Committee Tells China', *Times*, 1 April, available from www.timesonline.co.uk/tol/news/world/asia/article3659665.ece (retrieved 10 July 2010).
McCahill, M. (2002) *The Surveillance Web: the Rise of Visual Surveillance in an English City*, Cullompton: Willan.
McCahill, M. and Norris, C. (2004) 'CCTV in London', Working Paper No. 6, from the *Urban Eye: On the Threshold to Urban Panopticon? Analysing the Employment of CCTV in European Cities and Assessing its Social and Political Impacts*, European Commission project, available from www.urbaneye.net/results/ue_wp6.pdf (retrieved 1 February 2010).
Macpherson, W. (1999) *The Stephen Lawrence Inquiry: Report of an Inquiry by Sir William Macpherson of Cluny*, London: HMSO.
Magnier, M. (2008) *Beijing Olympics Visitors to Come Under Widespread Surveillance*, Los Angeles Times, 7 August.
Marsh, P., Fox, K., Carnibella, G., McCann, J. and Marsh, J. (1996) *Football Violence in Europe*, Oxford: Social Issues Research Centre.
Marx, G.T. (1988) *Undercover: Police Surveillance in America*, Berkeley: University of California Press.
Mathiesen, T. (1997) 'The Viewer Society: Michael Foucault's "Panopticon" Revisited', *Theoretical Criminology*, 1(2): 215–234.
Metropolitan Police (1999) *Operation Marathon: the Investigation*, available from www.met.police.uk/news/stories/copeland/thehunt.htm (retrieved 1 July 2010).
Mills, D. (2006) 'The Case for Shared National Requirements: Olympic Security', paper given at *Technology for Security and Resilience* conference hosted by the Homeland

Security Department of the Royal United Services Institute for Defence and Security Studies, London, 26–27 June.

NACRO (2002) *To CCTV or Not to CCTV: a Review of Current Research into the Effectiveness of CCTV Systems in Reducing Crime*, Community Safety Practice Briefing, May.

Norris, C. (2006) 'Closed Circuit Television: a Review of its Development and its Implications for Privacy', paper prepared for the Department of Homeland Security Data Privacy and Integrity Advisory Committee quarterly meeting, 7 June, San Francisco, CA.

Norris, C. and Armstrong, G. (1999) *The Maximum Surveillance Society*, Oxford: Berg.

Olympic Delivery Authority (2007) *ODA Security Industry Day, Call for Security Tenders*, available from www.london2012.com/documents/oda-industry-days/oda-security-industry-day-presentation.pdf (retrieved 3 December 2008).

Parkins, G. (2007) 'The CCTV Effectiveness Review and National Strategy', paper given at the *Local Government Association 2007 CCTV Conference: The Development of a National Strategy, Innovative Systems, Effectiveness and Standards*, Local Government House, London, 4 July.

Pecora, V. (2002) 'The Culture of Surveillance', *Qualitative Sociology*, 25(3): 345–358.

Peng, J. and Yu, Y. (2008) 'Beijing Olympics Security Plan', paper presented at *Security and Surveillance at Mega Sport Events: from Beijing 2008 to London 2012* conference, 25 April.

Reiner, R. (2007) *Law and Order: an Honest Citizen's Guide to Crime and Control*, Cambridge: Polity.

Samatas, M. (2007) 'Security and Surveillance in the Athens 2004 Olympics: Some Lessons from a Troubled Story', *International Criminal Justice Review*, 17(3): 220–238.

Sanan, G. (1996a) *Olympic Security 1972–1996: Threat, Response and International Co-Operation*, unpublished PhD Thesis, University of St Andrews.

Sanan, G. (1996b) 'Olympic Security Operations 1972–94', in A. Thompson (ed.) *Terrorism and the 2000 Olympics*, Sydney: Australian Defence Force Academy.

Scarman, L. (1981) *The Scarman Report: the Brixton Disorders, 10–12 April 1981*, London: HMSO.

Seabrook, T. and Wattis, L. (2001) 'The Techno-Flâneur: Tele-Erotic Re-Presentation of Women's Life Spaces', in L. Keeble and B. Loader (eds) *Community Informatics: Shaping Computer-Mediated Social Relations*, London: Routledge.

Security Products (2007) *Beijing To Invest More Than $720 Million For 2008 Summer Olympic Games Security*, available from http://secprodonline.com/articles/2007/08/14/beijing-to-invest.aspx.

Sherman J. and Ahmed, M. (2008) CCTV and DNA Advances Add to Bills but Minister Calls Rises Unacceptable, *Times*, 28 February, available from www.timesonline.co.uk/tol/news/politics/article3448837.ece (retrieved 2 May 2010).

Shute, S. (2007) *Satellite Tracking Of Offenders: a Study of the Pilots In England and Wales*, Ministry of Justice Research Report, London: HMSO.

Smith, G.J.D. (2007) 'Exploring Relations Between Watchers and Watched in Control(led) Systems: Strategies and Tactics', *Surveillance and Society*, 4(4): 280–313.

Tilley, N. (1993) *Understanding Car Parks, Crime and CCTV: Evaluating Lessons From Safer Cities*, Home Office Police Research Group Crime Prevention Unit Series Paper no. 42.

Tilley, N. (1998) 'Evaluating the Effectiveness of CCTV Schemes', in C. Norris, J.

Moran and G. Armstrong (eds) *Surveillance, Closed Circuit Television and Social Control*, Aldershot: Ashgate, pp. 139–175.

Toohey, K. and Veal, A. (2007) *The Olympic Games: a Social Science Perspective*, Wallingford: Cabi.

Webster, C.W.R. (2004) 'The Evolving Diffusion, Regulation and Governance of Closed Circuit Television in the UK', *Surveillance and Society*, 2(2/3): 230–250.

Wells, A. (2007) 'The July 2005 Bombings', paper given at the *Local; Government Association 2007 CCTV Conference: The Development of a National Strategy, Innovative Systems, Effectiveness and Standards*, Local Government House, London, 4 July.

Welsh, B.C. and Farrington, D.P. (2002) *Crime Prevention Effects of Closed Circuit Television: a Systematic Review*, Home Office Research Study 252, London: Home Office, available from www.homeoffice.gov.uk/rds/pdfs2/hors252.pdf (retrieved 1 February 2010).

Williams, C. (2003) 'Police Surveillance and the Emergence of CCTV in the 1960s', *Crime Prevention and Community Safety*, 5(3): 27–38.

Wilson, J. (2002) *Street Cameras Defended Despite Limited Effect Claim*, Guardian, 29 June, p. 6.

7 Strategic security planning and the resilient design of Olympic sites

Jon Coaffee

Introduction

The International Olympic Committee (IOC) makes explicit in guidance to host cities that it is *their* responsibility to provide a safe environment for the 'Olympic Family' (competitors, officials and dignitaries), while ensuring that such securitisation does not get in the way of the sporting activities or 'spirit' of the Games. As Thompson (1999: 106) observed, 'the IOC has made clear that the Olympics are an international sporting event, not an international security event, and while Olympic security must be comprehensive it must also be unobtrusive'. However, given the changing nature of the terrorist threat, 'securing' the Olympics is increasingly difficult and costly to achieve (Coaffee and Wood, 2006). Certainly, with Olympic planning being predicated on the 'severe' level of terrorist threat (Lord West, 2009), there is the possibility of seeing core notions of Olympic spectacle, to some extent, replaced by dystopian images of 'cities under siege' as organisers and security personnel attempt to deliver an Olympics in maximum safety and with minimum disruption to the schedule (Coaffee and Johnston, 2007; Boyle and Haggerty, 2009). In turn, the deployment of mass security generates Olympic 'spaces of exception' that have become standardised, mobile and globalised, and have the potential to become disassociated from the specific geographical contexts of host cities (Coaffee and Fussey, 2010).

As has been well documented, as a result of increased fears of international terrorism catalysed by the events of 11 September 2001, the cost of security operations surrounding major sporting events, in particular the summer Olympic Games, has increased dramatically. Since the 2004 Games in Athens, a large proportion of this increased cost can be attributed to extra security personnel as well as an array of temporary security measures, especially those which are effective at stopping or minimising the impact of vehicle borne improvised explosive devices (VBIEDs) (Coaffee and Johnston, 2007). For most Olympic organisers, their preparations for the Games necessarily include attempts to equate spectacle with safety and to 'design out' terrorism, often by relying on highly militarised tactics and expensive, detailed contingency planning (Coaffee, 2009).

The Olympic Games have become an iconic terrorist target, which imposes a burden of security on host cities well beyond what they would otherwise face

(Fussey et al., 2010). Security planning is now a key requirement of bids submitted to the IOC by prospective host cities, and has become a crucial factor in planning the Games. There is now a broadly accepted security management model for Olympic Games which is partially readjusted according to local circumstances of place. This includes elements of governance and organisation which seek to forge relationship between the numerous public safety agencies and the local Organising Committee. Here, as Jennings and Lodge (2009) highlight, for Olympic-type events, security arrangements tend to layer over existing national and international infrastructures (or at least those components that fit the standardised security framework). As well as the planned deployment of police and military personnel, and the co-option of private security and safety volunteers, this security regime also deploys the latest technology in an attempt to plan for and deter terrorist attack. For example, recent summer Olympiads (as well as other major sporting and cultural events) have become highly militaristic at certain geographical locations through the construction of large-scale bunkers and barriers, secure fencing around the key sites, as well as almost ubiquitous closed-circuit television camera coverage. Pre-Games and Games-time monitoring of key sites also commonly occurs through the use of, for example, scuba divers and helicopters, and an array of high-tech surveillance and monitoring devices. Integrated real time technologies are also now commonly deployed to aid the security effort, but with varying degrees of success (see, for example, Samatas, 2007).

Such exorbitant levels of security thus temporarily transform the host cityscape into a series of temporary exceptional spaces with displacement of the law by authority as special regulatory regimes are brought to bear so as to control behaviour and maintain order (Agamben, 2005). As Browning (2000) observed during the period immediately before the 2000 Games, 'Sydney in September will be under siege'. Likewise, for the duration of the 2004 Games, Athens became a 'panoptic fortress' to give assurances to the rest of the world that the city was safe and secure to host the world's greatest sporting spectacle (Samatas, 2004: 115).

Securing Olympic spectacle and protecting against the dangers of terrorist violence remains the *overriding* concern for Olympic organisers. This has become increasingly stark in the United Kingdom as London gears up for hosting the 2012 Games. Security concerns and responses not only played a critical part in the bidding process, but have also dominated media discussion after the host city was announced. This was especially the case after the series of terrorist bomb attacks on the London transport network on 7 July 2005 – the day after London was confirmed as host city. These attacks have led to a massive increase in perceived security needs and prompted organisers to draw up ever-more detailed plans. At the time of writing this has led to the initial security bill escalating from £224 million to over £600 million (plus a further £237 million 'contingency' fund), the planned adoption of advanced biometric security systems to monitor crowds, officials and athletes, special 'measures' to track suspects across the city; and, of particular importance to this chapter, a series of efforts to 'design-in' counter-terrorism features to the physical infrastructure of the Olympic venues and environs.

As briefly highlighted above, in recent summer Olympiads a host of *temporary* security solutions have literally swamped the host cities and in particular the sporting venues. For example, in 2004 the Olympic stadium in Athens, always likely to constitute the most spectacular target, received the heaviest fortification. According to Peek (2004: 6) the main stadium in Athens was 'supposed to be one of the most secure places on earth, impenetrable to terrorists plotting a possible attack on this summer's Olympics'. However, *permanent* design and architectural features intended to counter such threats have, to date, been largely absent from Olympic security preparation, only becoming a key strategic issue during the ongoing redevelopment of sites and venues for the 2012 summer Games in London. This interest in how planners, architects, developers and urban designers, alongside security specialists, can design-out terrorism (or, more correctly, design in counter-terrorism features) for Olympic facilities in London is a function of not just the supposed threats faced against 'crowded places' but also the longer term regeneration vision for the areas in the post-Games period, where issues of safety and security will be paramount within 'legacy' community facilities (Coaffee, 2009). Such permanence of security infrastructure also features prominently in Rio de Janeiro's successful candidate file for the 2016 summer Games.

The paradox of defending crowded places

Traditional approaches to crime prevention through design modification (most notably Secure by Design or Crime Prevention Through Environmental Design) are being increasingly hybridised with military security planning to provide protection against explosions at 'at risk' sites. Such attempts to reduce terrorist risk are by no means unprecedented in the UK. During the 1990s, the experience of UK authorities in attempting to 'design out' terrorism was largely confined to efforts to stop car bombing by the Provisional Irish Republican Army (PIRA) against the economic infrastructure in London (Coaffee, 2003). Before the events of 11 September 2001, threats of international terrorism predominantly came from VBIEDs targeting major financial or political centres. In response, attempts to counter terrorism often utilised planning regulations and advanced technology to create 'security zones' or 'rings of steel' where access was restricted and surveillance significantly enhanced (Coaffee, 2004; Coaffee *et al.*, 2008). September 11th and subsequent no-warning attacks, often using person-borne improvised explosive devices (PBIEDs) aimed at causing mass casualties, made such counter-terrorist tactics appear inadequate, and security policy began to shift to proactive and pre-emptive solutions often involving changes in the way in which design solutions could be unobtrusively constructed and space could be managed.

More recently, in the mid 2000s, concerns about the likelihood and impact of terrorist attacks on crowded public places using a combination of traditional and increasingly innovative methods has heightened the sense of fear in many urban locations, with future attacks against 'soft targets' appearing more likely (Coaffee, 2008). This was especially evident in light of the suicide attacks on the London transport network in July 2005 and the London and Glasgow attacks in July 2007 targeting a nightclub and airport respectively. Crowded public places (e.g. shopping areas, transport systems, sports and conference arenas) in particular are deemed to be at high risk (Coaffee et al., 2008). Furthermore, in most cases, such spaces cannot be subject to traditional security approaches such as searches and checkpoints without radically changing public experience and creating 'spaces of exception'. In the case of major event planning, such as the Olympic Games, the requirement to create more resilient urban areas that are less likely to suffer attack through 'designing in' effective counter-terrorism should, wherever possible, be done in a way that is as 'invisible' as possible to avoid creating unnecessary fear (ibid.).

National policy-makers and the Security Services now perceive attacks against crowded public places as 'inevitable', and securing such locations has thus become one of their key priorities in the ongoing fight against terrorism (Coaffee and Bosher, 2008). As UK Prime Minister Gordon Brown noted in a 'statement on security' (25 July 2007):

> The protection and resilience of our major infrastructure and crowded places requires continuous vigilance. I can confirm that over 900 shopping centres, sports stadiums and venues where people congregate have been assessed by counter-terrorism security advisers and over 10 000 premises given updated security advice.[1]

Concern is also being expressed about management of the security threat, in particular security procedures and queuing at airports and other facilities at which rigorous security checks are in force. These queues may themselves become targets since they are outside secure areas and could easily be attacked. This later point, in particular, represents a challenge at major sporting events such as the Olympics where queuing for admission to specific venues will be widespread. Together, this blend of changing terror methods and targets, especially those directed at crowded places in urban areas and around sporting venues, are providing challenges for security professionals and practitioners.

In the UK, enhancing security in crowded places has been backed up by large and ongoing streams of work being conducted by the National Counter Terrorism Security Office (NaCTSO) on disseminating protective security advice to places deemed vulnerable to targeting (including sports stadia and major events), and which calls for counter-terrorism measures to be embedded within the design, planning and construction of public places. At present across London there are no regulations in place to insist that developers, planners or architects regularly even consider counter-terrorism measures when designing a new development or public space. That said, at the time of writing, the UK Home Office is developing a counter-terrorism supplement linked to its existing planning guide,

Safer Places: the Planning System and Crime Prevention (Home Office, 2004) which, when adopted, will become a practical guide (but not regulation) for planners to consider how they can 'design in' counter terrorism, particularly in crowded locations.

Defending the crowded spaces of the Olympics

In previous summer Olympics, security in and around the stadium has generally been comprehensive and overt. Barcelona 1992 saw the deployment of over 25,000 security personnel due to fears expressed over reprisal terror attacks linked to the recently finished 1991 Gulf War, the fear of possible ETA attack and other localised threats. The result was a highly militaristic security regime at certain sites including the main stadium. The 1996 Games in Atlanta, held in the wake of the attempted bombing of the World Trade Center in New York in 1993 and the 1995 Oklahoma City attack, led to increased attention being paid to the protection of buildings. Although such attacks were neither entirely unprecedented nor unanticipated, they were significant in the mainstreaming of counter-terrorism features within the US planning system to protect both individual buildings and commercial districts (Coaffee and O'Hare, 2008). Such alteration in the built form was referred to by some at the time as the 'architecture of paranoia' (Brown, 1995) as a result of the practical response by statutory agencies who adopted crude but robust approaches to territorial security as their key modus operandi. By extension, the potential role that urban built-environment professionals could play in 'terror proofing' cities became more apparent, although such approaches were not considered for the Atlanta Games for a number of reasons – including cost and the fact that new venues were already in construction in the early 1990s, meaning that designed-in solutions, best done at the concept stage of the design process, were not feasible. Rather than using permanent design solutions, the strategy was to 'police out' terrorism using 20,000 military and law-enforcement personnel and 5,000 unarmed volunteer security personnel, and was 'Hi-tech and measured'. This latter objective involved the field-testing of NASA technology in civilian areas (Coaffee and Johnston, 2007). Security planning also 'tried to cover every angle possible – from cops on patrol to scuba divers and helicopters and high tech devices such as ID badges with computer chips' (Macko, 1996).

In the 2000 Sydney Games, although the cost and sophistication of security rose steeply from 1996, a key objective of securitisation was that security was designed to be omnipresent but unobtrusive, although in practice the swamping of venues and surroundings with security personnel and associated searching procedures was still ubiquitous. In the first post-September 11th Games in Athens, security, and its cost, were intensified. A five-fold increase in expenditure and a massive deployment of over 100,000 police and military personnel was intended to deter a possible attack. Such overt policing of space was backed up by an ever advancing array of security hardware, including a network of 13,000 surveillance cameras, mobile surveillance vans, chemical detectors, a number of Patriot anti-aircraft missile sites, NATO troops specialising in weapons of mass destruction, AWACS early warning surveillance planes, police helicopters, fighter jets, minesweepers and monitoring airships. Significantly, the Olympics forced the Greek state to speed up the modernisation of its state security system (Coaffee and Johnston, 2007). For the duration of the Games, Athens became a 'panoptic fortress' to give assurances to the rest of the world that the city was safe and secure to host the world's greatest sporting spectacle (Samatas, 2004: 115). However, the retrofitting of such security systems was envisioned as a long-term infrastructure project that was to be maintained after the Olympics and which critics have argued will become a menace to privacy and civil liberties (ibid., 117; see also Coaffee and Wood, 2006; Boyle and Haggerty, 2009).

What we can gauge from a study of securitisation utilised by host Olympic cities (and other major events such as football Word Cups) are a series of normalised event-security features which combine temporary physical features and the officious management of spaces with the aim of projecting an air of safety and security, both for visitors as well as potential investors (Coaffee *et al.*, 2008).

First, there is intense pre-planning, involving the development of control zones around the site, procedures to deal with evacuation, contamination and decontamination, and major incident access. Technical information is also scrutinised for all structures and ventures so that any weaknesses and vulnerabilities can be planned-out in advance. Second, there is the development of 'island security' involving the 'locking down' of strategic (vulnerable) areas of host cities with large expanses of steel fencing and concrete blocks surrounding the sporting venues (see, for example, Coaffee and Rogers, 2008). This combines with a high visibility police presence, backed up by private security, the Security Services, and a vast array of permanent and temporary CCTV cameras and airport-style checkpoints to screen spectators. Often events such as the Olympics are used to field test 'new' technologies. Third, to back up the intense 'island security', peripheral buffer zones are often set up in advance containing a significant visible police presence. This is commonly backed up by the presence of law enforcement tactics such as police helicopters, a blanket 'no fly zone', fleets of mobile CCTV vehicles, road checks and stop and search procedures. The result of these measures is that, often, access to 'public' spaces is restricted on public roads and on public footpaths, as a result of 'security concerns' (Coaffee *et al.*, 2008). Fourth, the enhanced resilience that such security planning, in theory, delivers is actively utilised as a future selling point for urban competitiveness in that the ability to host such an event in a safe and secure fashion, and without incident, is of significant importance in attracting future cultural activities and in branding a city as a major events venue. Fifth, there is increased evidence from major sporting events that a lasting benefit of hosting events is the opportunity for the retrofitting of permanent security infrastructure linked to longer term crime reduction strategies and 'legacy' (Coaffee and Wood, 2006).

Furthermore, bidding to host the Olympics is also, in many cases, considered a strong enough stimulus to develop robust and de facto permanent security planning procedures. For example, the recent attempt by Cape Town, South Africa, to bid (unsuccessfully) for the 2004 Games required the city to be seen as secure enough to host the Olympics. As a result, an extensive security infrastructure was introduced into areas posited as likely venues and visitor accommodation centres (Minnaar, 2006). After their bid failed, the CCTV systems were not removed; instead, they were justified as part of a general programme to combat crime, which in this case was seen as discouraging foreign tourists, investors and conference delegates (Coaffee and Wood, 2006).

Defending the main stadium and other Olympic venues

The iconic site for each summer Olympiad is its main stadia which becomes an inherently political space and hence a terrorist target for the duration of the Games. Its architecture is emblematic of the current state of modernity and reflects the political realities and aspirations of state leaders, as perhaps best articulated by the Olympic stadiums in Berlin 1936, Montreal in 1976 and Beijing in 2008. Architecture, and the built form more generally, has the capacity to transmit a range of dominant ideologies and emotions, potentially illustrating how a particular society is materially inscribed into space and how the built form possesses the power to condition new forms of subjectivity and everyday public experiences (Coaffee *et al.*, 2009).

The usual model of Olympic security planning, as noted in previous sections, focuses on restricting access and regulating space in and around key facilities. Until the upcoming Games in London in 2012, little had been done to utilise urban design principles and resilient materials to mitigate the risk of vehicle bombs in host cities. In the build up to 2012, and given the ongoing 'war on terror' and the requirements of a safe legacy environment, design has become a key component of security preparedness in and around the main Olympic site in Stratford, East London, where a host of new facilities are being built.

Securing East London

For effective security planning it is clear that solutions need to combine managerial coordination *and* innovative design approaches to physical security, to ensure that London, and in particular the Olympic Park area containing the main venues, does not become 'siege like' but resilient to possible attack. In this sense, counter-terrorist security must be comprehensive, but also as unobtrusive as possible (Coaffee, 2009). As the then Head of 2012 Olympic Security, Tarique Ghaffur, noted in 2007:

> This is a celebration of what London is about and of the Olympics... It's not about security or safety. Making the games as accessible as we can without security being obtrusive, is the trick we have to pull off.
>
> (Cited in Culf, 2007: 15)

Indeed, London 2012 (2004: 27) argued in their bidding documents that surveillance and security operations will begin at the start of construction and involve adaptation for every venue and will continue throughout the Olympic Games.

In terms of stadia construction, it is clear that 'Secure by Design' approaches will be utilised as they have been for a number of recent London developments, including Heathrow Terminal Five, the Millennium Dome, Wembley and Lord's Cricket Ground. This will involve the embedding, at the concept stage, of design features such as access control and integrated CCTV, the designing-in of 'stand-off areas' for hostile vehicle mitigation, as well as the use of more resilient building materials.

Around the main venues in Stratford there has also been talk of setting up advanced screening access points – the so-called 'tunnel of truth' which can check large numbers of people simultaneously for explosives, weapons and biohazards and could utilise face-recognition CCTV comparing visitors against an image store of known or suspected terrorists (Coaffee, 2009). Such an approach might also serve to restrict the need for queuing and hence reduce potential vulnerabilities.

Such a security approach is quickly becoming standard practice in the UK which has seen potential terror attacks against stadiums, and moreover their crowds, elevated to the top of the 'at risk' list. In 2007, the UK's National Counter Terrorism Security Office (NaCTSO), the police unit responsible for providing guidance in relation to business continuity, designing out vehicle borne terrorism, the protection of crowded places, and reducing opportunities for terrorism through environmental design, released a specialist guide for stadium developers, owners and operators, called *Counter Terrorism Protective Security Advice for Stadia and Arenas* (NaCTSO, 2007). Released in the wake of the July 2005 attacks against the London transport network, this guide set out how stadium management and design might be modified, or designed-in, to restrict the opportunities for terrorism to occur, as well as to aid evacuation and swift recovery should an attack be successful.

In this context, security planning and design of the stadium is required to take into account the terrorist threat and devise appropriate exterior security and vehicle management within any new building or renovation work. This will allow 'counter terrorism specifications, e.g. concerning glazing and physical barriers to be factored in, taking into account any planning, safety and fire regulations' (p. 1). The aim here is to 'put in place security measures to remove or reduce vulnerabilities to as low as reasonably practicable bearing in mind the need to consider safety as a priority at all times' (p. 11). Of particular concern within the NaCTSO guide, and within Olympic preparation, has been the threat posed by VBIEDs or car bombs. As the guide notes:

> They are capable of delivering a large quantity of explosives to a target and can cause a great deal of damage. Once assembled, the bomb can be delivered at a time of the terrorist's choosing and with reasonable precision, depending on defences. It can be detonated from a safe distance using a

timer or remote control, or can be detonated on the spot by a suicide bomber. Building a VBIED requires a significant investment of time, resources and expertise. Because of this, terrorists will seek to obtain the maximum impact for their investment. They generally choose high profile targets where they can cause the most damage, inflict mass casualties or attract widespread publicity.

(p. 37)

The guide (p. 37, amended) goes on to state that stadium developers and managers can do a variety of things to protect against VBIED:

- Ensure basic good housekeeping such as vehicle access controls and parking restrictions. Do not allow unchecked vehicles to park next to or under your stadium.
- Consider using physical barriers to keep all but authorised vehicles at a safe distance. Seek the advice of your local police Counter Terrorism Security Adviser (CTSA) on what these should be and on further measures such as electronic surveillance including Automatic Number Plate Recognition (ANPR) and protection from flying glass.
- Insist that vehicles permitted to approach your stadium are authorised in advance, searched and accompanied throughout. The identity of the driver should be cleared in advance.
- Do what you can to make your stadium blast resistant, paying particular attention to windows. Have the stadium reviewed by a qualified security engineer when seeking advice on protected spaces, communications, announcement systems and protected areas.
- Establish and rehearse bomb threat and evacuation drills. Bear in mind that, depending on where the suspected VBIED is parked and the design of your building, it may be safer in windowless corridors or basements than outside.
- Assembly areas must take account of the proximity to the potential threat. You should bear in mind that a vehicle bomb delivered into your building – for instance via underground car parks or through the front of your premises – could have a far greater destructive effect on the structure than an externally detonated device.
- Train and exercise your staff in identifying suspect vehicles, and in receiving and acting upon bomb warnings. Key information and telephone numbers should be prominently displayed and readily available.

Adapted from *Counter Terrorism Protective Security Advice for Stadia and Arenas* (NaCTSO, 2007: 37)

The stadium-specific security is strategically aligned with that of the Olympic Park in which the facility sits – a large, expansive area that is undergoing securitisation in preparation for 2012. The site for the main Olympic Park in Stratford, East London, has already been partially securitised. It was 'sealed' in July 2007 and nearby public footpaths and waterways closed for public access. The 11-mile blue perimeter fence – 'cordon blue' which was put in place for 'health and safety' reasons – has been likened by some to the Belfast peace walls (Beckett, 2007). This was subsequently replaced in large part by electric security fencing in late 2009. Biometric checks are routinely carried out on the construction workforce within the sealed site. What is clear, though, is that security at Games time will be ratcheted up significantly with an undoubtedly imposing and visibly policed security cordon encircling the site, whilst in the Olympic Park (inside the cordon), landscaped security and crime reduction features, infrastructure strengthening (for example bridges and other structures), and electronic devices that scan for explosives, are being embedded in order to push threats away from the Olympic site. Likewise, concealment points – areas where explosives might be hidden – are being scrutinised and, where possible, removed (for example, bird boxes or litter bins) or sealed (for example, drains).

These design features, if implemented, would be designed to be as unobtrusive as possible, and with a view to being kept in place post-Games for legacy purposes in order to deter activities such as joyriding, ram-raiding, drug-dealing, prostitution and general anti social behaviour. This means that features that are designed in specifically for counter terrorist purposes must, wherever possible, have a crime prevention capability in the legacy period when the park will be fully open to the public. Embedded safety and security features must therefore also be as 'invisible' as possible and be *proportional* to the threat faced.

Physical security will also need to be strategic and extend beyond the Olympic Park sites, as fears of the 'displacement', to other key sites in London, of possible terrorist attack persists. As such, London authorities will be using advanced surveillance to track suspects across the city, including London's ever-expanding system of Automatic Number Plate Recognition cameras (Fussey, 2007). This is seen as a soft touch approach, and preferable to having a police officer on every corner. An extra 9,000 officers are expected to be on duty in London at peak time, although concern has been expressed about leaving other parts of the UK vulnerable to attack if police officers are drafted in from other forces. As Tarique Ghaffur, the then Head of Olympic Security, noted in 2007, Olympic security is a pan-London operation:

> The whole rhythm of life in London will change as a result of these events and for 60 days we will have to take charge of that and make it safe in a way that people can enjoy themselves ... [an extra] 9000 officers at the peak is a heck of an ask.
>
> (Cited in Culf, 2007: 15)

In the build-up to the Games, different scenario table-top tests will increasingly be played-out for dealing with major incidents, including terrorism, to allow logistics such as cordon placement and evacuation routes to be planned in advance. The aim of all the security preparations and testing is to allow 'customer sensitive' security to prevail, which will provide the highest possible levels of security without resulting in having to 'lock down' major parts of the city, as has happened with other Olympiads (for example, Seoul, Sydney and Beijing, see Coaffee and Fussey, 2010). This will apply to the sporting venues as well as other key locations such as transport hubs and tourist areas.

Securing a safe legacy

At present, the security planning for the 2012 Olympics is ongoing and beginning to focus its efforts upon design challenges as part of a broader Olympic Safety and Security programme. This, in particular, will focus upon reducing the vulnerability of important national infrastructures and crowded places in order to 'protect the London 2012 Games Venues, events, transport infrastructure, athletes, spectators, officials and other staff and ensure their safe enjoyment of the Games' (HM government, 2009: 13). This is in line with the wider objective as outlined in the *London 2012 Olympic and Paralympic Safety and Security Strategy* published in July 2009: 'To host an inspirational, safe and inclusive Olympic and Paralympic Games and leave a sustainable legacy for London and the UK' (Home Office, 2009a). Moreover, within this strategy,

> safety encompasses the building of the Olympic venues and the placement of the overlay, for example, safety barriers, and the management of crowds within Olympic venues including those relevant aspects of the transport infrastructure and the management of crowds queuing to enter venues. Security encompasses the measures taken to mitigate the identified threats during the build, test, overlay and operational phases and the protection of persons using the venues in Games time, or queuing to enter them, and of the Games themselves.
>
> (p. 8)

The longer-term strategic regeneration of East London has also been a strong consideration when seeking to design in security at the Olympic venues. The principles on the accredited Association of Chief Police Officers' 'Secure by Design' scheme are being applied to venue construction as well as the 'white space' – the space in-between venues in the wider Olympic park area – 'in order to minimise crime and security risks' (Home Office, 2009b). It is also likely that, post-Games, the whole Olympic park will be granted Secure by Design status. This, it is hoped, will ensure that a safe and sustainable legacy is delivered to local communities, who will see benefits such as increased resilience to crime and anti-social behaviour in the parkland areas, as well as increased physical robustness being designed into sporting facilities for use by local people.

Similarly, Rio's successful candidacy to host the 2016 Olympic Games also draws on such 'secure by design' ideas (referred to here as 'CPTED', or 'crime prevention through environmental design'). To mitigate these risks (and those from general crime), a familiar plan is being developed to that in London:

A comprehensive security overlay will be implemented to ensure the integrity of all Games facilities and prevent unauthorized access. This will include perimeter security, integrated access control and alarm management, coupled with technical surveillance... The security overlay will be based on Crime Prevention through Environmental Design (CPTED) principles, to be incorporated into the design of all venues.

(Rio2016, 2007: 27)

It is clear that, in recent years, given the changing nature of terrorist tactics and targeting, that design-based and permanent solutions are being sought that aim to maximise long term (legacy) benefits of the security infrastructures that are implanted in order to secure the Olympics Games, which is a mobile and temporary event. In this regard, although London 2012 is in many ways following a traditional security model utilised in previous Games, it is also providing a new template for designed in security in and around the venue areas. This is already a practice that is being taken up by other candidate cities in their attempts to argue that 'their' city can deliver an Olympic spectacle in relative safety.

Note

1 See www.number10.gov.uk/output/Page12675.asp (accessed 4 May 2008).

References

Agamben, G. (2005) *State of Exception*. Chicago: University of Chicago Press.
Beckett, A. (2007) 'Cordon Blue', *Guardian*, 21 September, www.guardian.co.uk/society/2007/sep/21/communities.
Boyle, P. and Haggerty, K. (2009) 'Spectacular Security: Mega-Events and the Security Complex', *International Political Sociology*, 3: 257–274.
Brown, G. (2007) *Statement on Security*, www.number10.gov.uk/output/Page12675.asp (accessed 4 May 2008).
Brown, P.L. (1995) 'Designs in a Land of Bombs and Guns', *New York Times*, 28 May, p. 10.
Browning, M. (2000) 'Olympics Under the Gun', *Guardian*, 8 March, www.cpa.org.au/gachive2/991games.html (accessed 1 April 2000).
Coaffee, J. (2003) *Terrorism, Risk and the City*, Aldershot: Ashgate.
Coaffee, J. (2004) 'Rings of Steel, Rings of Concrete and Rings of Confidence: Designing Out Terrorism in Central London Pre and Post 9/11', *International Journal of Urban and Regional Research*, 28, 1: 201–211.
Coaffee, J. (2008) 'Redesigning Counter-Terrorism for Soft Targets', *Homeland Security and Resilience Monitor*, 7, 2: 16–17.
Coaffee, J. (2009) *Terrorism, Risk and the Global City – Towards Urban Resilience*, Aldershot: Ashgate.
Coaffee, J. and Bosher, L. (2008) 'Integrating Counter-Terrorist Resilience into Sustainability', *Proceeding of the Institute of Civil Engineers: Urban Design and Planning*, 161, Issue DP4: 75–84.
Coaffee, J. and Fussey, P. (2010) 'Security and the threat of Terrorism', in J. Gold and M.

Gold (eds) *Olympic Cities: City Agendas, Planning, and the World's Games, 1896 to 2012* (Second Edition), London: Routledge (forthcoming).

Coaffee, J. and Johnston, L. (2007) 'Accommodating the Spectacle', in J. Gold and M. Gold (eds) *Olympic Cities: City Agendas, Planning, and the World's Games, 1896 to 2012*, London: Routledge.

Coaffee, J. and O'Hare, P. (2008) 'Urban Resilience and National Security: the Role for Planners', *Proceeding of the Institute of Civil Engineers: Urban Design and Planning*, 161, Issue DP4: 171–182.

Coaffee, J. and Rogers, P. (2008) 'Rebordering the City for New Security Challenges: From Counter Terrorism to Community Resilience', *Space and Polity*, 12, 2: 101–118.

Coaffee, J. and Wood, D. (2006) 'Security is Coming Home – Rethinking Scale and Constructing Resilience in the Global Urban Response to Terrorist Risk', *International Relations*, 20, 4: 503–517.

Coaffee, J., Moore, C., Fletcher, D. and Bosher, L. (2008) 'Resilient Design for Community Safety & Terror-Resistant Cities', *Proceeding of the Institute of Civil Engineers: Municipal Engineer*, 161, ME2: 103–110.

Coaffee, J., O'Hare, P. and Hawkesworth, M. (2009) 'The Visibility of (In)security: the Aesthetics of Planning Urban Defences Against Terrorism', *Security Dialogue*, 40: 489–511.

Culf, A. (2007) 'Capital Will Need 9,000 Officers a Day to Police the Olympics', *Guardian*, 17 March, 15.

Fussey, P. (2007) 'Observing Potentiality in the Global City: Surveillance and Counterterrorism in London', *International Criminal Justice Review*, 17, 3: 171–192.

Fussey, P., Coaffee, J., Armstrong, G. and Hobbs, D. (2010) *Sustaining and Securing the Olympic City: Reconfiguring London for 2012 and Beyond*, Ashgate: Farnham (forthcoming).

HM government (2009) *Countering the Terrorist Threat: How Industry and Academia Can Play Their Part*, London: HMSO.

Home Office (2004) *Safer Places: The Planning System and Crime Prevention*, London: Home Office.

Home Office (2009a) *London 2012 Olympic and Paralympic Safety and Security Strategy*, London: HMSO.

Home Office (2009b) *A Safe and Secure Games For All*, London: HMSO.

Jennings, W. and Lodge, M. (2009) 'Governing Mega-Events: Tools of Security Risk Management for the London 2012 Olympic Games and FIFA 2006 World Cup in Germany', in *Proceedings of the Political Studies Association Conference*, 2009.

London 2012 (2004) 'Candidate File', www.london2012.com/documents/candidate-files/theme-12-security.pdf (accessed 12 December 2005).

Macko, S. (1996) 'Security at the Summer Olympic Games is Ready', *EmergencyNet News Service*, Vol. 2–191.

Minnaar, A. (2006) 'Crime Prevention/Crime Control Surveillance: Public Closed Circuit Television (CCTV) in South African Central Business Districts (CBDs)', paper presented at the *Crime, Justice and Surveillance* conference, Sheffield, 5–6 April.

NaCTSO (2007) *Counter Terrorism Protective Security Advice for Stadia and Arenas*, London: NaCTSO.

Peek, L. (2004) 'How I Strolled into the Heart of the Games', *The Times*, 14 May, p.6.

Rio2016 (2007) *Rio de Janeiro Applicant File – Theme 13: Security*, Rio de Janeiro: Brazil.

Samatas, M. (2004) *Surveillance in Greece – from Anticommunist to Consumer Surveillance*, New York: Pella Publishing.

Samatas, M. (2007) 'Security and Surveillance in the Athens 2004 Olympics: Some Lessons from a Troubled Story', *International Criminal Justice Review*, 17, 3: 220–238.

Thompson, A. (1999) Security, in R. Cashman and A. Hughes (eds) *Staging the Olympics – the Events and its Impact*. Sydney: University of New South Wales Press, pp. 106–120.

West, Lord (2009) 'Keynote Address: Olympic and Paralympic Safety and Security', paper presented at *Royal United Services Institute Olympic and Paralympic Safety and Security Conference*, RUSI, Whitehall, 13 November.

Part III
Coordination, roles and responsibilities

8 Governing the Games in an age of uncertainty
The Olympics and organisational responses to risk

Will Jennings

> Security and risk management are part of the Olympic package – as well as for almost every national and international gathering today.
> (Richard W. Pound, IOC Vice-President, 1996–2000)[1]

Risk, modernity and control

Some argue that the world has entered a new era of extreme events and risks (e.g. Beck 1992; OECD 2003; Lagadec and Michel-Kerjan 2005; Lagadec 2007), citing examples such as the AIDS epidemic; Chernobyl nuclear disaster in 1986; the Sarin nerve gas attacks on the Tokyo Subway in 1995; 11 September 2001 terror attacks; the major power outages in North America during August 2003; the Asian tsunami of December 2004; Hurricane Katrina in September 2005; and the Credit Crunch of 2008/2009. Beck (1992: 21) argues risk can be understood as modern society's response to 'hazards and insecurities induced and introduced by modernization itself'. This changing risk environment is attributed to processes of globalisation (in its economic, technological, cultural and environmental forms), increasing interdependence, urbanisation and technological innovation.[2] The scale and form of modern life is at the root of its own instability and vulnerability. The complex and systemic relationship between modern organisations, technologies and individuals is said to be a source of 'normal accidents' (Perrow 1984, 1999), exacerbated through failures of organisational culture (Vaughan 1997), with positive feedback processes that give rise to the social amplification of risk (Kasperson *et al.* 1988; Pidgeon *et al.* 2003). Whether or not such accounts are right to stress the newness of the risks encountered in modernity, they nevertheless shed light upon social, technological, economic and organisational productions of risk.

In this age of uncertainty, societies, economies and governments are increasingly organised in response to risk (Giddens 1991, 1999; Beck 1992), reflecting modernist aspirations of quantification, measurement and control (e.g. Porter 1995; Scott 1998). The idea of risk, Giddens argues (1999: 3), is interlinked with attempts to control the future. It acquired increasing influence in public and private institutions (see Hacking 1990; Bernstein 1996) after the emergence of probabilistic thinking during the nineteenth century (Hacking 1975). While risk is now ubiquitous, formal controls have also become pervasive in both government and business, in attempts to account for and manage risk (e.g. Hood *et al.* 1992, 1999, 2001; Hood and Jones 1996; Power 1997, 2004, 2007). This has coincided with wider shifts towards regulation as a mode of governing (e.g. Majone 1994, 1996; Moran 2003), and transformation of organisation of the modern state and its relations with its citizens and private interests. All these approaches to organisation share a concern with securing of order and control through instruments of surveillance, formal reporting, policing and enforcement. Risk is increasingly an organising concept (O'Malley 2004; Power 2007), in fields such as counter terrorism, nuclear energy, public health, financial markets and transport. The replication of organisational responses to risk across domains has been reinforced through the professionalisation of risk analysis as a generic practice since the late 1960s (Hacking 2003).

This chapter argues that changes in organisational responses to risk associated with the Olympic Games must be understood in the context of broader social, governmental, technological and economic trends, of which the rise of regulation and risk management in both the public and private spheres is just one recent phase. The analysis that follows offers a theoretical dissection of different forms of Olympic risk, describes a number of decision-making biases and errors that are common in the organisation of the Games, and reviews the changing nature of organisational responses to risk throughout Olympic history, with particular reference to security. It notes the origin of Olympic governance in entrepreneurial, philanthropic and municipal forms, moving towards state-dominated arrangements from the 1930s onwards, and since the 1970s increased transfer of risk to markets, increased dependence upon technological solutions, spread of regulation and risk-management as organising logics and methods, and the evermore interconnected character of information and organisational responses to risk. While the main bodies engaged in staging and securing the Olympics – the host government, the International Olympic Committee (IOC) and the host OCOG – are much the same as was the case in 1896, these shifts in the ownership and management of risk have contributed to a more complex and diverse organisational environment that, in turn, has had profound effects upon administration of the Games and its response to security and other risks.

The Olympics and risk

Since the revival of the modern Games by Baron Pierre de Coubertin at Athens in 1896, the Olympic Games and Olympic movement have encountered hazards and threats in a range of forms. Such risks have taken the form of boycotts, budget over-spends, terrorist incidents, operational failures, public-relations disasters, corruption scandals, refereeing protests, doping controversies, timekeeping problems and ambush marketing. Some of these risks threatened the long-term viability of the Games (Payne 2006: 5–12). Even Olympics on a smaller scale, such as the inaugural Games in Athens in 1896, Paris in 1900 and

St Louis in 1904, were afflicted by logistical difficulties (e.g. Guttmann 1992; Young 1996). The Games of 1908 were relocated from Rome to London (British Olympic Council 1908: 19) after the eruption of Mount Vesuvius (Guttmann 1992: 28–29), illustrating the risk of extreme geographical events to the organisation and staging of the Olympic Games. One of the longstanding concerns in organisation of the Games is the maintenance of effective financial controls; with budget over-spends experienced, for example, in Athens 1896 and 2004, London 1908 and 2012, Montreal 1976 and Sydney 2000. Many risks associated with governing the Games have been ever-present since its inception.

Since the Munich massacre at the 1972 Games, during which 11 Israeli athletes and coaches were murdered by a group of Palestinian terrorists, the Olympics have been a recognised target for terrorist threats. The events at Munich were broadcast live to a worldwide audience on television as the crisis unfolded, highlighting the global platform the Games offers for protests and attacks. Planning for Sydney 2000 proved prophetic of the Al Qaeda attack on 11 September 2001, as organisers prepared strategies for the scenario feared by IOC President, Juan Antonio Samaranch: that of a commercial plane being hijacked and flown into the opening ceremony (Magnay 2005). The bombing of a public concert at Centennial Park during Atlanta 1996, which killed two and injured 111 people, did not entail the same international dimension as Munich, but illustrates the potential threat of lone domestic attackers at the Games. Security in a broader geopolitical sense has been a risk for the Olympics ever since the cancellation of the 1916, 1940 and 1944 Games due to the outbreak of the First and Second World Wars. Military and territorial conflicts have, at times, affected national participation in the Games, as well as levels of security provisions required. Although perceived as a tit for tat response to the US-led boycott of the Moscow 1980 Games, the Soviet Union used the pretext of security concerns in withdrawal of its team from Los Angeles 1984.

The Olympics are both a magnet and an amplifier of organisational and operational risks. The event itself increases the probability and consequence of hazards and threats, at the same time as generating its own unique set of risks. Furthermore, its organisation encounters difficulties of calculating and managing risk that are a general feature of such mega-projects (Flyvbjerg *et al.* 2002, 2003; Altshuler and Luberoff 2003). Organisation of the Games is a vast and complex exercise undertaken under the global spotlight. It has been described as 'the world's largest peacetime event' (Higgins 2007), equivalent to the synchronous staging of 33 world championships in the same city. For example, the Beijing 2008 Olympic Games hosted 28 sports, with 302 events at 37 competition venues over 16 days of competition, bringing together 204 participating NOCs, 11,000 athletes, 5,500 officials and coaches, 2,500 referees and judges, 20,500 media, 70,000 volunteers and 4 million spectators, with an estimated global television audience of 4.7 billion people.[3] The event has been transformed in size and complexity from the 241 amateur athletes who competed in nine sports in the Games at Athens in 1896. An extensive programme of sporting competition now runs in parallel, dependent upon a large infrastructure network, water and power supplies, policing and security, emergency services, accommodation, ticketing and merchandise, broadcast and media communications, catering, financial transaction networks, procurement and IT systems. Risks and contingencies associated with organisation of the Games therefore encompass a wide range of natural hazards, man made hazards and man made threats. Such risks can take the form of environmental phenomena such as heatwaves or storms, organisational failures, network disruptions or terrorist attacks.

The megalithic proportions of Olympic infrastructure and operations are combined with a rigid schedule of ceremonial and sporting events. While the host government and OCOG, bound by the Host City Contract with the IOC, have no legal rights of cancellation, in practice implications of an event capable of prompting outright cancellation would in all likelihood be quite catastrophic and have global repercussions. Even at the height of the Munich crisis in 1972, following a one day suspension of events, IOC President Avery Brundage declared that 'the Games must go on' (Guttmann 1984: 140). The Games are subject to extreme costs associated with critical failures or incidents.

Two dimensions of Olympic risk

It is possible to differentiate between the sorts of risks encountered in the organisation and staging of the Olympic Games. Some highlight geopolitical, organisational, economic and reputational categories of risk (e.g. Jennings 2008; Pound 2009). Others explore the different organisational 'recipes' used to manage risk and crisis situations at the Games (e.g. Jennings and Lodge 2010, forthcoming). The risks associated with organising the Olympics take numerous forms and are located at different locations (e.g. the main Olympic site, urban centres, commercial districts, and transport networks and hubs). To understand risk as an organising concept, however, it is first important to understand both its probabilistic relationship to staging of the Games and the location of either its origin (i.e. cause) or its jurisdiction (i.e. effect). An alternative analytical approach to existing studies, offered in this chapter, seeks to determine whether the likelihood of particular hazards or threats is exogenous or endogenous to the event itself, and whether these risks are located at the national or sub-national level with the host government and organising committee, or at the transnational level with bodies such as the IOC or international sports federations. This provides insight both into the hazards and threats associated with the Olympics, as well as the changing organisational task of hosting the Games and of managing risk.

With this purpose, the following analysis makes the theoretical distinction between an internal–external dimension and a national–transnational dimension of Olympic risk. This is illustrated in Table 8.1. The internal–external dimension refers to whether risks are directly produced in organisation and staging of the Games (i.e. hazards or threats that are probabilistically dependent upon occurrence of the event) or whether instead they occur external to the event but nevertheless have effects upon it (i.e. hazards or threats that are probabilistically independent of the event, but that can disrupt its staging). This dimension captures, in essence, whether the Olympic Games either produces or is the venue for hazards and threats. The national–transnational dimension refers both to the location of the risk in terms of its underlying cause and the territory or jurisdiction in which responsibility for its management resides. Some risks are localised to a particular city or country, whereas others can spread across international boundaries and multiple territorial units. Likewise, responsibility for managing certain risks can reside with the host government and the host OCOG, and for others lie with the IOC and transnational bodies such as the IOC, NATO or the World Anti-Doping Agency. This dimension refers both to the national and sub-national levels because cross-national differences in political systems mean that organisation of the Games can be delegated to national, regional or municipal government (or a mixture thereof). These two cross-cutting dimensions are not exhaustive, but nevertheless provide a systematic framework for analysis of risks associated with organisation of the Games that also allows for integration of insights on the broader state of the risk environment. The four possible combinations of these two dimensions are now discussed.

Internal–national

The scale and the scope of the modern Olympics is a direct source of internal risks encountered at the national or sub-national level by the host city and host government. Growth over time in the total number of events, sports, athletes, officials, media and television viewers (for example) has increased the level of resourcing required to stage the Games. While risks of scale tend to be assumed at the national level, they are – at least in part – a function of the schedule of sporting events set by the IOC and its long-term expansion of the Olympic programme. The historical upward trend in most of the measures of event size is

Table 8.1 Two dimensions of Olympic risk

	Internal (endogenous)	*External (exogenous)*
Sub-national/national (host)	• Project management • Construction • Spectator flows • Completion deadlines • Revenue • Public support	• Urbanisation • Domestic terrorism and public disorder • Random events (e.g. weather, breakdowns) • Water and power supplies • Transport networks
Transnational	• Inter-games learning • Geo-politics • Doping • Exchange rates • Commodity Prices • Governance of the Olympic movement and international sport	• International terrorism and cyber-terrorism • Technological change • Global economic conditions (e.g. Credit Crunch) • Infectious diseases (e.g. avian flu) • War

illustrated in Figure 8.1. This increase in scale of the Olympics over time and increases in the revenues and costs of both the Games and the Olympic movement (Preuss 2000, 2004) means that there are greater economic risks attached for the host city and the IOC (i.e. potential losses increase proportional to the potential gains). While operational costs have tended to be offset against commercial revenue at recent Olympics, the costs of infrastructure, policing and other public services are typically underwritten by the host government. The bid, design, organisation and operation of the Games are increasingly complex due to the proliferation in the number of variables and decision points that are subject to uncertainty and, therefore, risk. This direct scale to risk relationship is present in most organisational and operational risks for the host OCOG and Olympic-related agencies (e.g. construction, personnel, procurement and event coordination). As the Games grows in size, so too do the number of opportunities for failure and the potential consequences of failure. Other local characteristics of the Games can generate project risks. The legacy promises of the London 2012 Olympic programme – as pledged in bid documents to the IOC – present a long-term delivery risk, while environmental aspects of the Nagano 1998 and Sydney 2000 Olympics created similar risks for organisers in fulfilling bid commitments. A sizeable proportion of risks associated with the Games are therefore 'self inflicted' inasmuch as they are inherent to the growth of the event and programme. This growth in scale and global audience also makes the Games an attractive target for (external) terror attacks, the eruption of geo-political tensions and other high-impact incidents (more of which in due course).

External–national

At the national and sub-national level, local geographical, economic and political factors external to planning for the Games are a potential source of Olympic hazards and threats. The geography and urban layout of the host city, along with the critical Olympic infrastructure (e.g. transport networks, venues and stadiums, water and power supplies, accommodation, media facilities) provides the underlying architecture for delivering and staging the Games. The local structure of public-transport links and road networks have consequences for the operation of ticket barriers, crowd control, policing, first aid, traffic management, catering and officiating of events. All of these geographical aspects structure both the effect and response to incidents such as a stadium fire or a terror attack. The efficient functioning of the local transport network is essential for spectator and competitor travel. So too is the resilience of electricity supplies, given the disruptive effect of power outages (see US–Canada Power System Outage Task Force 2004). Other unpredictable local events, such as extreme weather conditions (e.g. Klinenberg 2002; Lagadec 2004) also represent risks for staging the Games. Such events can occur either during or outside competition-time. Weather is a function of local context, generating different sorts of meteorological risk. In the run-up to the Sydney 2000 and Athens 2004 Games, for example, there was concern over the health of athletes competing in high temperatures. In

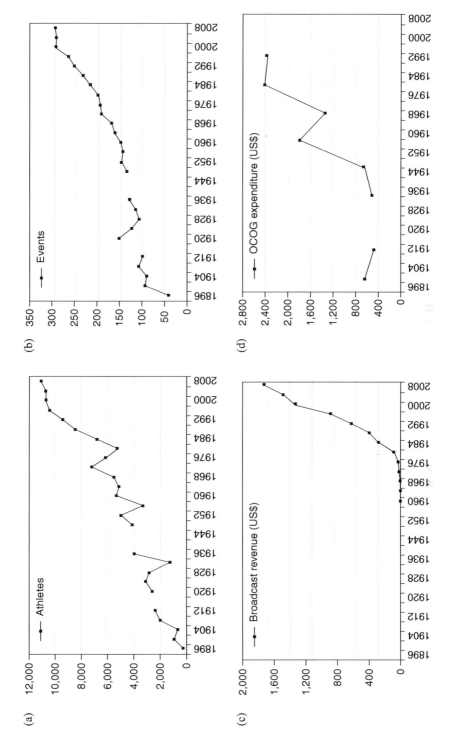

Figure 8.1 Selected indicators of organisational growth of the Olympic Games 1896–2008.

contrast, planners for Beijing 2008 were concerned with rainstorms and the health impacts of smog and pollution. Organisers' concerns during the run-up to the Vancouver 2010 Winter Olympics focused upon whether there would be enough, rather than too much, snow! The effects of such external events are not specific to the Games, but are significant because their impact would be amplified during Games-time due to the potential effect upon competition and spectators. It is possible that there might too be risks for local populations during competition time as a side effect of the diversion of emergency, disaster relief and public health agencies from their normal operations.

The probability of other risks at the national level might also be amplified due to the Games. For example, effects of a pandemic outbreak (e.g. Barry 2004) could be amplified by increased population movements during competition time. The unpredictability of power demand during mega-events such as the Olympics likewise increases the risk of outages, which can have knock-on effects on transport and venues (for example). In the run up to the Sydney 2000 Olympics, Kingsford Smith Airport suffered a power failure that created a backlog of flights – prompting concerns over its readiness for the traffic associated with the Games. In July 2004, a power outage across Athens and southern Greece left the Transport Minister, Mihalis Liapis, stranded in the middle of showcasing a new Olympic rail link to the airport. There may be high alert for other sub-national or national threats during the hosting of the Olympics. The prominence of the Games can provide a platform for dissidents, civic protests, and terror groups and individuals (such as Basque separatist incidents prior to the Barcelona 1992 Olympics and the bombing of the Centennial Park during Atlanta 1996). It can also stimulate protest (such as human-rights protests about Beijing 2008) and public disorder (such as riots prior to the Mexico City 1968 Olympics). These risks use the Olympics as a target, encouraged by its profile, but are not in themselves created by the event.

Internal–transnational

The transnational character of the organisation of the modern Olympics means that it is exposed to risks that cross both international boundaries and Games. The demand for raw materials for use in construction of Olympic venues and infrastructure introduces risk of fluctuations in commodity prices and in supply volumes. This is a longstanding risk of stadium construction, with post-war shortages of steel a problem for organisers of the 1948 London Olympics (Cabinet minutes 1947: 226). Likewise, the dependence of the IOC and host OCOG upon commercial revenues from broadcast rights adds the risk of exchange-rate movements, with the largest share of revenues paid in US$. Fluctuation in value of the Canadian dollar against the US dollar led to a $150 million shortfall in the financing model of the Vancouver 2010 Winter Olympics, with a loss of revenue from broadcast rights and international sponsorship revenues (see Auditor-General of British Columbia 2006). Transnational governance of the Olympic movement and international sport also entails risk, such as the effective transfer of knowledge between events and problems such as doping and corruption. Allegations over corruption of IOC delegates and organisers in the Sydney 2000 and Salt Lake City 2002 bids led to controversy and the resignation of numerous officials. Some operational risks, such as the broadcasting of information in real time to the world through Internet and television networks, exist at a transnational level. Such risks are not specific to Games hosted in a particular location or at a particular time but, instead, are part of the wider Olympic risk environment.

The threat of international terrorism is perhaps the most prominent and recurring internal–transnational risk affecting Olympic organisation, brought first to prominence with events in Munich in 1972 and again highlighted with the Al Qaeda attacks on the US on 11 September 2001. There are strong transnational elements to security aspects of the Games. Historically, the Games has also often been vulnerable to diplomatic incidents and geo-politics (see Espy 1979; Hill 1997), such as the Berlin 1936 'Nazi Olympics', the Cold War boycotts of the Moscow 1980 and Los Angeles 1984 Olympics, and protests in the run-up to Beijing 2008 over China's human-rights record. These risks are external to the organisation of the Games itself but are, nevertheless, an intrinsic part of the Olympic package. The Games therefore encounters some risks that are present in all contexts and across international borders.

External–transnational

Last, there are some events or circumstances with potential consequences for the Olympics that are probabilistically independent of it. The 11 September 2001 attacks had significant impact on planning for subsequent Games, inflating the security costs and encouraging the involvement of transnational defence and intelligence organisations such as NATO. Likewise, changes in global economic conditions can impact upon local planning for the Games. The Great Depression reduced participation in the Los Angeles 1932 Olympics, while post-war austerity measures affected the design of the London 1948 Olympics. The recent Credit Crunch made real the identified risk of securing private finance for capital projects for the London 2012 Olympics (National Audit Office 2007: 16) as well as increasing the likelihood of insolvency of Olympic suppliers and contractors. Another source of risk external to the Games and the Olympic movement is technology, which nevertheless has been associated with dramatic change in the organisation and staging of the Games as well as its broadcasting. Increased technological innovation also increases the number of things that can go wrong. Extreme events such as international crises, war or infectious disease can also interrupt the Games. The World Wars resulted in suspension of the four-year cycle of the Olympics, with no Games held in 1916, 1940 or 1944. A global outbreak of infectious disease, such as that recently feared of avian flu, would likewise have serious repercussions for organisation of the Games in its effects on public health as well as on border controls.

It is evident, then, that threats and hazards at the Games come in numerous forms. Both the probabilities and the prospective losses which together constitute

Olympic risk can be endogenous (internal) or exogenous (external) to the event and can reside at the national or sub-national level with the host government and organising committee or at the transnational level with international bodies such as the IOC. The internal dimension of risk indicates that both probabilities and losses can be amplified by the act of staging the Games itself – with increases in scale also resulting in increases in complexity and potential losses (as well as potential gains). While external risks are independent – in probabilistic terms – these again are amplified through scale. At the same time, certain risks are a product of features particular to the host city or host government, whereas others reflect broader social, technological political and economic processes occurring at the transnational level. Binary presentations of categories of risk (e.g. Jennings 2008) do not reveal everything about the complex and interdependent risk environment in which Olympic organisation and operations occur. The following analysis introduces some of the decision-making biases and errors that are common to organising the Games and that can distort the handling of some of these risks.

Optimism bias, discounting, risk-aversion and normal accidents

Organisation of the Games is susceptible to a number of decision-making biases and organisational effects that can accentuate existing hazards and threats or structure the responses of organizers to them (see Jennings and Lodge forthcoming). There is, on the one hand, a taste for risk in the act of bidding for the right to host the Games and in perceptions of the controllability of such a vast enterprise. At the same time, there is considerable risk aversion from host governments and organising committees, as well as the IOC, which is reflected in hyper-active efforts to mitigate and manage risk. This reflects more general tension inherent in the grand ambitions of 'high modernist' projects and the risk-averse and controlling aspirations of regulation as a mode of governing (e.g. Scott 1998; Moran 2003). The following discussion notes, in brief, effects on risk-management of optimism bias, risk discounting, risk aversion and the 'normal accidents' (Perrow 1984, 1999) that arise from complex organisational systems.

Optimism bias

Organisation of the Olympics is often associated with organisational and operational processes that encourage 'optimism bias' concerning both risks and benefits. Optimism bias (e.g. Flyvbjerg *et al.* 2002, 2003) refers to systematic over-estimation of likelihood of positive outcomes and, therefore, underestimations of risk. Because the right to stage the Games is awarded through a competitive candidature procedure, Olympic bids are often designed to win votes of IOC members and tend to underemphasise inadequacies (Luckes 1997). This creates divergence between the formal bid documents – described by former IOC Vice-President Dick Pound as the 'most beautiful fiction' (Pound 2004) – and subsequent programmes of infrastructure construction and event operations. This gap is further amplified through the relatively small proportion of resources assigned to bid formulation and reliance upon templates rather than planning from first principles, which tends to produce generic bids rather than encourage the systematic challenging of planning assumptions. Emphasis on previous 'bid documents' (see Luckes 1998) and a 'specimen bid' (Arup 2002a) during preparation of the London bid for the 2012 Olympic Games are typical of the discounting of risk during the candidature phase and relative powerlessness of the IOC to prevent budget inflation given the extended lag time of the process. The final cost of the Games often exceeds initial expectations. The most famous case was the $1 billion deficit incurred at Montreal 1976, but Games have often incurred substantial deficits. Most recently, organisation of the Vancouver 2010 Winter Olympics and the London 2012 Summer Olympics have incurred cost over runs despite concerted attempts at technical forecasting and control of expenditure (Auditor General of British Columbia 2003; National Audit Office 2007).

Risk discounting

Optimism bias during the candidature process can be amplified by uncertainties in discounting of risk. This arises from inherent difficulties of predicting future outcomes. The extended time duration of the bid, planning and organisation of the Games, quite often between 15 and 20 years in gestation, causes a high level of uncertainty in the forecasting of variables such as infrastructure costs, revenue streams and security threats. As a result, the width of confidence intervals around predictions becomes larger over time. Such errors can arise through gradual divergence from the initial expectation (e.g. creeping cost inflation) or sudden and exceptional changes in planning assumptions due to extreme events or complete revisions of risk perceptions (e.g. terrorist incidents, global or domestic economic shocks). Consider the example of the London 2012 Olympic Games. The genesis of the London bid originated in unsuccessful bids from Manchester for the 1996 and 2000 Games that were formulated during the early 1990s. However, even since formulation of the initial feasibility studies for a London bid (Luckes 1997, 1998; British Olympic Association 2000), the national and transnational risk environment has undergone fundamental changes such as the increased terrorist threat from Al Qaeda and extremist Islamic groups and the decreased threat in the British mainland from Irish Republicanism after the Good Friday Agreement of 1998 and the process of devolution to Stormont. Likewise, warnings about the threat of an avian flu pandemic (alongside the swine flu pandemic of 2009) are contingencies that were unanticipated in the formulation of the bid. Most of all, organisers have been required to respond to increased financial risk due to the global financial crisis, in terms of pressure on public sector expenditure and decreased investment and sponsorship from the private sector.

Risk aversion

Alongside features of Olympic organisation that tend to encourage underestimation of risk, the global profile and political status of the Games – making it a perpetual media event – can encourage a zero tolerance attitude towards potential threats and risks. This heightens organisers' attention to reputational and political factors, such as geo-political risks, which can in turn generate organisational or operational difficulties. After the Cold War boycotts of the 1980s, the IOC under the leadership of Juan Antonio Samaranch adopted a proactive approach to management of political risk associated with the Games (Payne 2006: 14). Host governments and organising committees are also sensitive to perception of the Games. After a series of minor operational mistakes and failures at the Atlanta 1996 Olympics led to it being nicknamed the 'glitch Games' (Sack 1996), a feasibility study for the London 2012 bid noted that 'the Atlanta experience showed the media can play an important role in defining the perception of the success or otherwise of the Games' (Luckes 1997: 66), promoting the importance of state of the art facilities for the Olympic broadcast and media centres. Organisation of the Games can also be shaped by more general aversion to risk within government. For example, intervention of federal government in security planning for the Salt Lake City 2002 Winter Olympics – designated as a National Special Security Event by the Office of Homeland Security – was characteristic of the heightened state of concern within the Bush Administration and increased influence of ideas such as Vice-President Cheney's 'one per cent doctrine' (Suskind 2006), which treated a one per cent chance of a terrorist threat as a certainty in terms of formulation of security responses.

Normal accidents

The exceptional scale and complex interdependence of Olympic infrastructure (e.g. transport networks, stadiums, water and electricity supplies, accommodation and media facilities) and operations (e.g. ticket barriers, policing, traffic management, catering, first aid, officiating of events), and time dependence of planning and organisation, increases the risk of unexpected interactions between multiple failures, errors and incidents. More often than not, the Games requires construction of new infrastructure and facilities, and escalation of routine levels of public and private sector activities such as air and rail transport services, policing and emergency services. This complex and interconnected web of organisation, operations and technology makes the Olympic Games vulnerable to the 'normal accidents' identified by Perrow (1984, 1999), for example in the potential for an isolated security incident to have unforeseen cascading effects on the functioning of energy and transport networks already operating above normal capacity. With events as well as the athletes' accommodation and press facilities concentrated at the main site for most recent Games, a single incident has the potential to have far-reaching effects across both the site and the programme of events. The effects of such incidents can also be amplified by the time dependence of organisation, with the immovable schedule of ceremonial and sporting events. Furthermore, bespoke design of Olympic infrastructure and facilities, all delivered just-in-time, limits opportunities for stress-testing and event rehearsals. While cost over-runs and delays are a defining characteristic of mega projects (Altshuler and Luberoff 2003; Flyvbjerg *et al.* 2002, 2003), there is little room for miscalculation in the construction programme for the Games.

The risks encountered in organisation of the Games, outlined earlier, are therefore subject to distortion, in both the under and over estimation of risk, as well as potential for the unexpected interaction of incidents across jurisdictions. Some of these biases and effects are reflected in organisational responses to risk. It is to this Olympic governance of risk that the discussion turns to next.

Organisational responses to risk at the Games

To understand the contemporary state of Olympic management of risk, it is essential to understand the historic forms that the administration of the Games has taken over more than a century. These organisational responses often draw upon the prevailing doctrines and methods in politics, business and public administration at the time. It is possible to distinguish distinct organisational eras from landmark Games, innovations or incidents, each of which resulted in lasting change. In the earliest period from the 1890s, during the modern revival of the Games and the Olympic movement, entrepreneurial, philanthropic and state-sponsored forms of organisation were most dominant. This era was followed by increasing state based coordination and control from the 1930s, as the Nazi Olympics of 1936 marked a watershed in direct state involvement in the Games that has to a large extent persisted to the present day. The active involvement of the state in responses to risk matched the more general growth in size of government through episodes of crisis (e.g. Higgs 1987). While national and sub-national governments remain influential – to a greater or lesser degree according to the national context – a number of distinct trends have occurred in Olympic governance of risk since the 1970s.

First of all, with increasing dependence of the IOC upon revenues from broadcasting and sponsorship contracts, and after a large financial deficit incurred at Montreal 1976, the economic impact – and risk – of the Olympics has become central to its organisation. Around this same period, the 1980s – and the Los Angeles 1984 Olympics in particular – marked the rise of commercial practices and increasing transfer of risk (and benefits) to the market. The use of joint public–private initiatives has been common since Barcelona 1992, and these changes reflect the wider trend of state retraction (Majone 1994, 1996) as well as increased influence of market-like values in public management (Hood 1991; Pollitt 1995). Since the late 1980s, the Olympic movement and the Olympic Games have been governed increasingly through logics and technologies of regulation, audit and risk-management – such as in the international harmonisation of standards and rules and in the use of formal internal controls within organisations. This again is symptomatic of wider trends towards regulation (e.g. Majone 1994; Hood *et al.*

1999; Moran 2003) and audit and risk-management (e.g. Power 1997, 2007) as modes of governing within public and private organisations. Since the late 1990s – and following the Sydney 2000 Olympics in particular – organisational responses to risk have become ever-more interconnected and standardised, with the replication of strategies and the emergence of a professional community of experts in Olympic bidding and management. This exhibits processes of institutional isomorphism (DiMaggio and Powell 1991), as risk is increasingly active as a 'boundary object' (Power 2004: 34) that enables conversation across interests, professions and organisational functions.

Last, and somewhat at odds with prevailing trends in organisation since the 1970s, the Games have in recent times become subject to processes of securitisation (Buzan *et al.* 1998), after the events of 11 September 2001, as escalated Olympic security programmes have deployed intelligence services and military forces and technologies to monitor and manage the threat from terrorism. This shift was first in evidence at the Salt Lake City 2002 Winter Olympics and has been evident ever since in the mushrooming security costs of the Games. The resurgence of state intervention in response to security risks is therefore just one of the more recent chapters in contemporary Olympic management of hazards and threats, in contrast to market oriented mechanisms that have become more popular in Olympic organisation since the 1970s.

These different eras or landmarks of organisational responses to risk are outlined in Table 8.2. While these organising authorities – the IOC, host government and organising committee – have performed a similar function for most Olympics since 1896, there has been a fundamental shift in how the Games are governed, with increasing reliance upon market based mechanisms, growth of internal controls and formal management of risk, and ever-more interconnected and standardised organisational forms. These broad trends are, of course, subject to variations due to national context. Nevertheless, they indicate long-term shifts in organisation of the Games and its response to threats and hazards. In the remainder of this chapter, these broad changes in organising logics are analysed in greater detail, including discussion of the management of security risks associated with the Olympics.

Table 8.2 Landmarks in organisational responses to risk at the Olympic Games

1896 (Athens): Entrepreneurialism, philanthropy, patronage and state sponsorship
1936 (Berlin): State-based coordination and control
1976 (Montreal): Economic evaluation and impact analysis
1984 (Los Angeles): Contracting out, commercialisation and new public management
1988 (Calgary): Regulation, audit and risk-management
2000 (Sydney): Transnational networks in regulation and risk-management
2002 (Salt Lake City): Securitisation and the resurgence of state intervention

From entrepreneurialism, philanthropy and state sponsors to state-led organisation

The limited size and profile of the earliest modern Olympics from the 1890s to the 1930s did not stimulate a great degree of formal attention to risk. During this period, organisers tended to be most concerned with securing financial, sporting and political support to stage the Games (e.g. MacAloon 1981; Young 1996). The Olympics were often dependent upon private contributions from influential individuals drawn from business, politics and nobility. For example, with the Greek Treasury in a state of bankruptcy, and its government 'staunch' in its denial of public funds, construction of the main stadium for the Athens 1896 Olympics was financed through a donation from the philanthropist, M. George Averoff (de Coubertin *et al.* 1897: 22). Plans for the Games to be held in Chicago in 1904 were abandoned so that it could be integrated into the staging of the more business-focused St Louis Exposition (Sullivan 1905: 159). Construction of the facilities for the London 1908 Olympics was, similarly, financed by the authorities organising the Franco-British Exhibition (British Olympic Council 1908: 26), although co-location of both the Exhibition and the Games at a single site also resulted in problems of crowd control and in securing of the stadium (ibid.: 391). The final report of the British Olympic Council (1908: 20) noted the 'well-known and generally accepted maxim of English life that undertakings such as these shall be carried out by private enterprise, and without help of any sort from the government'.

In some contrast to this era, the period from the 1930s onwards marked an increase in the active role of government in organising the Games. This reflected the growing size of the Olympics, in terms of the infrastructure requirements, financial commitment and potential national-security implications. State-based responses to risk are characteristic of most Games since the 1930s, as national or sub-national governments have assumed the role of 'backer of last resort'. In preparations for the Los Angeles 1932 Olympics, the Californian state legislature passed the California Olympiad Bond Act 1927, establishing the California Olympiad Commission as the organising authority and issuing one-million dollars-worth of state bonds to finance the event. While the Berlin (Nazi) 1936 Olympics marked a dark chapter in Olympic history, the German regime's high level of control over organisation also signalled the increasing involvement of government in staging the Games. The Nazi regime was active in the administration and policing of the Games, in particular through the *Deutscher Reichsbund für Leibesübungen* (the Reich Commission for Physical Training), providing both facilities and manpower to support the Organising Committee as well as managing and financing construction of the Olympic stadium. It also deployed military resources, with German armed forces undertaking 'police and patrol' service for the Olympic Village, in addition to the government's provision of fire and other policing services (Organisationskomitee für die XI. Olympiade Berlin 1936: 210). Despite the quite exceptional circumstances surrounding the Nazi Olympics, the influence of government in organisation of the Games remains an integral feature of Olympic governance.

Olympic Organising Committees (OCOGs) continue – more than 70 years later – to depend upon government, at either the national or sub-national level, to underwrite most of the financial risks associated with the Games. Government has often been left to resolve the problem of cities being left with post-Games deficits. The famous example is the $1 billion debt incurred by organisers of the Montreal 1976 Olympics, paid off thirty 30 years later through a tax surcharge on local residents, after its Mayor Jean Drapeau had declared that '[t]he Olympics can no more have a deficit than a man can have a baby'. The public sector contribution to the Sydney 2000 Olympics turned out to be six times greater than the original bid (NSW Audit Office 1999; UK Select Committee on Culture, Media and Sport 2003), while the total cost of the Athens 2004 Olympics escalated from £3.2 billion to £6.3 billion (House of Commons Library Research Paper 2005: 37). One notable exception to the requirement of government to provide guarantees of financial support was the Los Angeles 1984 Olympics. In 1978, the City of Los Angeles passed a voter-approved measure – Charter Amendment 'N' – that prohibited the expenditure of municipal funds on staging the Olympics without a guarantee of reimbursement. Special dispensation from the IOC enabled the Los Angeles Organising Committee to operate the Olympics as a commercial venture, generating a budget surplus of $232.5 million through commercial revenue and fund-raising (see LA84 Foundation 2004: 17).

Security costs outside the main Olympic site also tend to be absorbed by government, with changes in the global security context leading to heightened risk-management. For example, the Organising Committee for the Salt Lake City 2002 Winter Olympics received an additional $300 million from the US federal government to cover its extra costs after the terror attacks of 11 September 2001. The security costs of Athens 2004 also increased, from €515 million to €970 million, in response to this change in threat from international terrorism. There is, despite this, a general bias towards underestimation of security costs at recent Games, with the budget for the Vancouver 2010 Winter Olympics increasing from $175 million to $900 million, and from £190 million in the London bid for the 2012 Olympics to £600 million in later estimates. Host governments tend to finance the management of security risks that emerge external to the Games (see Table 8.1). Similar to the Berlin 1936 Olympics, government's provision of manpower from the armed forces and the police remains integral to securing most recent Games. At the Athens 2004 Olympics, around 70,000 police and armed forces were on patrol in central Athens and at Olympic venues, in addition to support from NATO and European Union forces. The number of police (15,000) to be deployed at the London 2012 Olympics is smaller, but nevertheless represents about 10 per cent of total UK police manpower.

Weakness of the state can itself contribute to organisational failures in managing risk. The West German authority's post-war lack of a specialist anti-terrorism unit has been attributed as one of the causes of the botched rescue of the Israeli hostages and escape of a number of the Palestinian terrorists at the Munich 1972 Olympics (Reeve 2000), as the constitution prohibited federal armed forces from operating inside Germany during peacetime. The Munich police and the Bavarian authorities responsible for security had no previous training in hostage crisis operations and lacked the appropriate equipment (see Schreiber 1972; Groussard 1975). As will be noted later, limits of state capacity can lead to networked and transnational approaches to the management of security risk.

Economic evaluation and transfers of risk to the market

Since the 1970s, organisation of the Games has combined an interest in its economic impact with increasing transfer of risk – and risk management – to the market. The first economic impact study (Iton 1978) was undertaken for the Montreal 1976 Olympics, and the evaluation of economic impacts has been repeated on a regular basis, such as for the Los Angeles 1984 Olympics (Economic Research Associates 1984) and the Sydney 2000 Olympics (PricewaterhouseCoopers 2002). The rise of economic impact analysis reflects growth in the potential financial risk and benefit – to both the host city and government and to the IOC – of staging the Games, as Olympic costs and revenues have increased in real terms (see Preuss 2004).

Alongside this concern with economic impact, a growing trend in organisation of the Games and in governance of the Olympic movement is the transfer of risk and benefits to the market, combined with the use of complex financial instruments. The contracting out of delivery and service functions to public/private or private organisations can range from catering to security. State-owned enterprises have been responsible for the construction of infrastructure and facilities in organisation of recent Olympics. The government-owned Barcelona Olympic Holding SA (HOLSA) constructed the main competition venues, new road infrastructure and the Olympic Village for the Barcelona 1992 Olympics, on behalf of the national and city government, financed with a mix of public and private funds. The Bird's Nest stadium for Beijing 2008 was developed by the China International Trust and Investment Corporation (CITIC) consortium and Beijing Municipal Government's Beijing State-Owned Assets Management company (BSAM). Other public bodies such as London 2012's Olympic Delivery Authority contract commercial partners (the CLM Consortium) to assume some responsibilities for project delivery. Such arrangements are designed to decentralise project risks, such as in procurement and cost-management.

This transfer of responsibilities to the private sector is also observed in the widespread use of security contractors alongside regular policing operations in securing the Games. Security contractors were used in large numbers at both Sydney 2000 (5,000 police and 4,500 contractors)[4] and Athens 2004 (25,000 police and 3,500 contractors), with similar levels in provisional estimates for London 2012 (15,000 police and 6,500 contractors) (see London 2012 bid, 2004: Volume 3, 39; Mayor of London 2004: Q. 357/2004).

Transfer of risk to the market can itself bring further risk, however. At the height of the global financial crisis, Lend Lease, developer of the Olympic Village for London 2012, experienced difficulty in securing private equity and debt funding for the project. It was unable to raise its financial commitment of £650 million, prompting a government-led rescue package that consisted of additional funding from its contingency fund, further public funding from the Homes and Communities Agency and over £150 million in loans from a banking consortium. The Sydney 2000 Olympics encountered similar problems in a shortfall of more than A$100 million against projected ticket and sponsorship revenue (NSW Parliament 2000), requiring contingency funding and prompting the New South Wales state government to assume responsibility for operational programmes and budgets (with a few exceptions) through the Olympic Coordination Authority.

Another approach to management of financial risk that transfers risk to the market is insurance. It has become prominent during the past decade as the IOC and Organising Committees have taken out insurance policies to cover organisational risks associated with security threats – in particular those posed by international terrorism. Protection of commercial interests in staging of the Games has been a concern for broadcasters since the 1970s. Although the US-led boycott of the Moscow 1980 Olympics forced US broadcaster NBC to cancel its coverage of the Games at a loss of $30 million, it was able to recoup some expenses from its insurance cover with Lloyd's of London. This has become standard practice for Olympic broadcasters exposed to increased financial risk due to the growth of their advertising revenue. NBC again purchased insurance cover for the Beijing 2008 Olympics. Because the host government is required, under the host city contract, to provide the IOC with a guarantee of financing for the Games, Olympic Organising Committees have not tended to take out insurance, although organisers of the Salt Lake City 2002 Winter Olympics purchased cancellation cover from Lloyd's of London even prior to the events of 11 September 2001.

Commercial revenues associated with the Games are the main source of income for the IOC, however, leading it to purchase insurance for cancellation due to terrorism or natural disaster for recent Games. It first purchased insurance cover for the Athens 2004 Olympics, with $170 million cover at a premium of $6.8 million (Buck 2004), later taking out $415 million cover at a premium of $9.38 million for the Beijing 2008 Olympics as part of a long-term policy that spans numerous Games (see Lenckus 2008). The IOC is vulnerable to interruption of its cyclical revenue stream and transfers risk to the market in response. Such arrangements are, however, designed to mitigate the after effects of security incidents rather than protect against occurrence of the incidents themselves.

The rise of regulation, audit and risk-management

Since the late 1980s and early 1990s, risk has been ubiquitous in governance of the Olympic movement and organisation of the Games – with increased oversight of Olympic bids and host preparations and in the spread of accounting, audit and risk-management practices in staging the Games (see Jennings 2008). The candidatures procedure is now highly standardised, with the IOC providing a template and questionnaire for bids in its IOC *Candidature Procedure and Questionnaire* (IOC 2004a). This documentation provides detailed specification of organising strategies, and is supported through assessments of the IOC Evaluation Commission, which report on the risks associated with each bid. For the 2012 Games, the selection process was a self-confessed exercise in risk-assessment, with the Evaluation Commission describing its task as 'a technical and fact-finding one: to verify the information stated in the candidature file, to determine whether proposed plans are feasible and to make a qualitative assessment of risk' (IOC 2004b: 5). After the award of the Games, regular monitoring and evaluation of organisational progress is conducted through the visits and reports of the IOC Coordination Commission.

Since the 1990s, the risk management mechanisms imported from the private sector have become increasingly complex, sometimes requiring management of second order risks. The risk transfer agreements negotiated with venue developers for the Vancouver 2010 Winter Olympics enabled the Organising Committee (VANOC) to minimise its level of capital expenditure, but required regular project and contract monitoring since non-completion was a risk created by transfer of delivery functions (see Auditor-General of British Columbia 2006). The management of risk therefore can create opportunity costs – at least on paper. To manage risk or not manage risk, that is the question. Inaction over implementation of a proposed hedging strategy resulted in a loss of around $150 million in broadcast and international sponsorship revenues for Vancouver 2010 due to a decline in strength of the Canadian dollar (Auditor-General of British Columbia 2006: 7). To similar effect, the London 2012 Organising Committee (LOCOG) wrote off £27 million on paper in its 2008 Annual Report (LOCOG 2008: 15) due to the difference between the hedged rate and the spot rate in its hedging contracts protecting against revenues paid in foreign currency (although further fluctuation in exchange rates could either improve or worsen this situation before 2012).

With the ongoing spread of risk-management practices in organisation of the Games, London 2012 perhaps might have a claim to be the first 'risk based' Olympics in terms of its organising principles and in the wide range of strategies put in place to manage and mitigate risks associated with delivery and staging of the Games across the whole of the Olympic programme – encompassing policy, infrastructure, security, finance and legacy functions (see National Audit Office 2007). Audit and the management of programme risks is conducted by the Olympic Board and Government Olympic Executive, informed by general information on threats and hazards (such as from the Cabinet Office's National Risk Register) or reports from the risk registers or risk logs of organisations responsible for delivery or operations such as the Olympic Delivery Authority or the Metropolitan Police's Olympic Security Directorate.

Transnational networks of regulation and risk-management

Over the past couple of decades, management of risk encountered in the organisation of the Games has acquired an ever more transnational and networked character – through the influence and interaction of international sports federations, anti-doping agencies, national security agencies from other countries and commercial interests. At the same time, emergence of new global hazards and threats has further encouraged networked approaches to the management of risk. Organisational responses to terror threats since 11 September 2001 have depended in equal measure upon policing manpower on the ground and intelligence-gathering, exchange and processing often across international borders and jurisdictions. Security responses to risk tend to either access existing transnational networks or lead to the creation of new structures, further increasing the geo-political aspect of the Olympics. The exceptional security situation associated with staging the Games can cause even superpowers such as the US or China to encounter information and resource shortages, prompting transnational cooperation and risk sharing.

The IOC and the host (national) government are often reliant upon assistance from foreign governments and transnational organisations to redress limits of resources or expertise. Exchange agreements in intelligence have been a feature of Olympics since the 1990s, such as Olympic Intelligence Centres established at Atlanta 1996, Sydney 2000 and Athens 2004. This delegation of security functions has become particularly common since the events of 11 September 2001. For the Athens 2004 Olympics, the Greek government received guidance and support from the US government and the NATO Alliance (and the US provided the International Atomic Energy Agency $500,000 for radiation detectors). Ahead of the Beijing 2008 Olympics, the US sent its Nuclear Emergency Support Team to provide assistance in response to concern over radiological threats, despite the absence of previous defence collaboration between the US and Chinese governments. This is not just a recent phenomenon. Ten days before the Mexico City 1968 Olympics, the Mexican government launched a crackdown on urban protesters leaving a considerable death toll. The US Central Intelligence Agency, which had been monitoring the unrest, recognised the weakness of the host government and sent it military radios, weapons, ammunition and provided riot control training before and during the Games (Doyle 2003).

This increasing influence of transnational organisational networks is also observed outside the defence and security domain, in relation to matters such as doping and the regulation of sport. The World Anti-Doping Agency (WADA) was established to regulate the abuse of drugs in international sport in November 1999 as an initiative of the IOC and, although an independent foundation, receives funding from the IOC and national governments. Its World Anti-Doping Code, implemented by sports authorities prior to the Athens 2004 Olympics, harmonises rules and regulations concerning doping across all sports and countries. This regulatory regime covers more than 500 sports authorities including International Sports Federations, the IOC and National Olympic Committees that have adopted the code. The code is subject to ongoing transnational consultation and revision, with the most recent version taking effect on 1 January 2009. It has been implemented through national governments' ratification of the UNESCO

International Treaty Against Doping in Sport, adopted by 191 governments in 2005 (effective February 2007). This depends upon sports authorities to enforce anti-doping regulations within their own jurisdictions outside Games time. An earlier transnational agreement in the regulation of sports doping was the Anti-Doping Convention of the Council of Europe in Strasbourg, established in 1989 which, like the WADA code, sought the harmonisation of standards and regulations related to doping and testing. While transnational regimes standardise management of doping risks, inequities in monitoring resources available to national federations or in sports with lower revenue streams create potential for the uneven enforcement of doping regulations – so that controls are, in fact, not all risk-based.

Another mechanism for the transnational spread of practices of risk-management in the organisation of the Olympics occurs through the contracting out of essential services in project management and delivery to major transnational corporations. In recent times, a small world of Olympic partner firms, contractors and consultants has become integral to organisation of the Games, enabling a degree of continuity between host cities. These transnational firms provide services for functions such as IT, communications, finance, engineering and planning, project management and security. This gives rise to a closed network of Olympic organisation in some areas, where an elite community of firms and experts are frequently contracted to multiple Games. It also reflects the small global pool of firms possessing both the capacity and expertise to support an event as large and as complex as the Olympics. For example, considerable technological functions are entailed in staging the Games – in particular, provision of support to results, information feeds and media facilities. Both the IOC and organising committees have insufficient expertise and organisational capacity to deliver these 'in-house'. At the Beijing 2008 Olympics, ATOS Origin, the IOC's Worldwide IT Partner for the Olympic Games, managed an IT Team of 4,500 personnel, supporting 10,000 computers, 1,000 servers, 200,000 accreditations, 4,800 result system terminals and 4,000 printers. Its services were operational 24/7, including implementation of a large-scale testing programme. ATOS also led technological project management for Salt Lake 2002 (then as SchlumbergerSema), Athens 2004, Turin 2006, Beijing 2008 and Vancouver 2010, and is contracted to London 2012.

Since the 1990s, a similar pattern of the transfer of organisational responsibilities to corporate firms and consultants is evident for delivery of venues and infrastructure at the Olympic Games. Arup – a global firm of consulting engineers, designers, planners and technical specialists – held the contract to provide transport planning for the Sydney Olympic Park, compiled analysis of parameters of a 'specimen bid' for the London 2012 bid (Arup 2002a) and was a member of consortiums responsible for delivering the Bird's Nest stadium and National Aquatics Center for the Beijing 2008 Olympics. Likewise, the CLM consortium (consisting of Laing O'Rourke, Mace and CH2M Hill) is contracted as the delivery partner for London's Olympic Delivery Authority. Of those firms, CH2M Hill provided programme management in preparation for the Atlanta

1996, Sydney 2000 and Beijing 2008 Olympics. The management of risk in organisation of the Olympics has therefore become more transnational and networked, and this in turn has contributed to isomorphism in the systems and technologies of risk-management implemented at the Games, as responses to security or project management risks have to some degree been replicated over time and across different settings.

Securitisation and responses to risk

The last of the distinct processes observable in organisation of the Olympics since the 1970s – and since 2001 in particular – is the increasing securitisation of the Games, as security responses have escalated in response to the threat of international terrorism. This has been evident in significant growth in security budgets for the Games since 2000 – as preoccupations with security have led to the under-estimation of security costs. The initial Arup study on the feasibility of a London bid included a 'provisional sum for the cost of all security for the Olympics following consultation with the Metropolitan Police and based on the experience of Sydney 2000 and Salt Lake City 2002' (Arup 2002a: 3–4), estimated at a cost of £160.2 million (Arup 2002b: 98). Even after the events of 9/11, consultants Arup reported that 'with more time to plan security for a 2012 Games, the costs are not likely to reach those incurred at Salt Lake City [£245 million]' (2002b: 95). In the London bid, site security was costed at £190 million, increasing to £268 million in the revised March 2007 budget which estimated the total security and policing cost at £600 million (House of Commons, Public Accounts Committee 2008a). Since the last official estimate, the reported security costs for London 2012 have reached £1.5 billion (e.g. Beard 2008). While direct comparison between Games is problematic, due to variations in the national governing context and security environment, it is clear that costs of securing the Olympics have been subject to considerable inflation over the past 30 years, and dramatically since the Sydney 2000 Olympics (see Roberts 2008).

Securitization of the Games has also led security responses to override preexisting institutional jurisdictions and measures. During the initial planning for the Salt Lake City 2002 Winter Olympics, a consortium of state, local and federal agencies (known as the 'Utah Olympic Public Safety Command') was delegated responsibility for management of all public safety activities. In 1999, the federal government's Office of Homeland Security intervened and designated the Games a National Special Security Event, which required the US Secret Service to take charge of security and made the FBI responsible for law-enforcement. The Salt Lake City security operations involved around 11,000 security personnel from more than 60 federal, state and local agencies such as the National Park Service (Rigg 2002). Securitisation of the Games has therefore observed a resurgence of state-dominated responses to risk, not only in the realm of security and defence but also in the interest of government in the organisation and operation of the Games.

Conclusion

This chapter has sought to demonstrate the importance of risk in organisation and staging of the Olympics, both in terms of the different forms in which risk is encountered and in change over time in organising strategies used for management and mitigation of risk. The analysis has shown how some risks are endogenous to the Olympics (i.e. these are subject to 'production' in the course of its organisation) while others are exogenous to it (i.e. these are independent of it in probabilistic terms, but its occurrence alters the potential losses of a particular event). Such risks can, furthermore, be distorted through biases and errors that are common to decision making in the organisation of the Olympics. In contrast to other studies, the chapter has demonstrated how changes in organisational responses to risk are not exceptional to the Olympics, but instead reflect fundamental shifts in the doctrines and practices of government and business. Since the 1970s, this is observed in organisers' increased concern with economic impact and transfer of risk to the market, alongside widespread use of regulation, audit and risk management as forms of external or internal control of organisations. Over time, similarities in organisational responses to risk exhibit some of the characteristics of isomorphism (see DiMaggio and Powell 1991) as practices of risk management are imported from one setting to another within the interconnected Olympic worlds of specialist firms and consultants. While the jurisdictions and responsibilities of the main authorities involved in organisation of the Games have changed little since those present at the earliest modern Olympics in 1896, these shifts in the management of risk have had deep rooted effects in changing how the Games is organised – both on paper and in practice.

Acknowledgements

The author thanks the Economic and Social Research Council for support through the ESRC Research Fellowship (Reference RES-063-27-0205), 'Going for Gold: The Olympics, Risk and Risk Management'.

Notes

1 Accessed online at www.business-in-asia.com/inside_olympics.html on 15 August 2005.
2 Systemic risk alone is not new. It is possible to identify numerous counter-examples suggesting that interdependence, scale and globalisation have been at work for centuries in exposing mankind to natural and man-made risk. Events such as the bubonic plague of the 1340s, Great Fire of London of 1666, Titanic disaster of 1912, Spanish flu outbreak of 1918 and the Great Mississippi flood of 1927 mirror many of the characteristics of their modern counterparts. Each of those historical events exhibit characteristics of extreme risks and failures of regulation and of risk-management. Certainly, extreme risk events today are more accelerated than their historical precursors (e.g. the spread of bubonic plague from Asia to Europe transpired over a period of years or decades, carried along trade routes, whereas the recent swine flu pandemic crossed the globe in a matter of weeks), and social and technological scale make the effects more pronounced, visible and measurable.

3 AGB Nielsen Media Research (www.nielsen.com).
4 See *Time Pacific*, 14 August 2000 (Number 32).

References

Altshuler, Alan A. and David Luberoff (2003). *Mega-Projects: The Changing Politics of Urban Public Investment*. Washington, DC: Brookings Institution.
ARUP (2002a). *London Olympics 2012: Costs and Benefits, Executive Summary.* 21 May. London: ARUP/Insignia Richard Ellis.
ARUP (2002b). *London Olympics 2012: Costs and Benefits*. Department of Culture, Media and Sport, Freedom of Information Request. London: ARUP/Insignia Richard Ellis.
Auditor-General of British Columbia (2003). *Review of Estimates Related to Vancouver's Bid to Stage the 2010 Olympic Winter Games and Paralympic Winter Games*. Victoria: Queen's Printer for British Columbia, www.bcauditor.com/pubs/2002–03/report6/OlympicGames.pdf.
Auditor-General of British Columbia (2006). *The 2010 Olympic and Paralympic Games: A Review of Estimates Related to the Province's Commitments*. Victoria, BC: Auditor General of British Columbia.
Barry, John, M. (2004). *The Great Influenza – the Epic Story of the Deadliest Plague in History*. New York: Penguin Books.
Beard, Matthew (2008). 'Security costs "will send 2012 bill over £10bn"', *Evening Standard*, 29 September.
Beck, Ulrich. (1992). *Risk Society: Towards a New Modernity*. London: Sage.
Bernstein, Peter L. (1996). *Against the Gods: the Remarkable Story of Risk*. New York: John Wiley & Sons.
British Olympic Association (2000). *London Olympic Bid: Confidential Draft Report to Government* (15 December).
British Olympic Council (1908). *Official Olympic Report*. London: British Olympic Association.
Buck, Graham (2004). 'Vaulting Olympic risk', *Risk & Insurance* (August), available at http://findarticles.com/p/articles/mi_m0BJK/is_9_15/ai_n6156490.
Buzan, Barry, Ole Waever and Jaap de Wilde (eds) (1998). *Security: A New Framework for Analysis*. Boulder: Lynne Rienner.
Cabinet minutes, CAB/128/9 (March 1947). Cabinet 33 (47). National Archives.
de Coubertin, Pierre, Timoleon J. Philemon, N.G. Politis and Charalambos Anninos (1897). *The Olympic Games, Part II*. Athens: Central Committee.
DiMaggio, Paul J. and Walter W. Powell. (1991). 'The Iron Cage revisited.' In Paul J. DiMaggio and Walter W. Powell (eds), *The New Institutionalism in Organisational Analysis*. Chicago: Chicago University Press.
Doyle, Kate (2003). 'The Tlatelolco Massacre: declassified U.S. documents on Mexico and the events of 1968', National Security Archive, George Washington University, 10 October, www.gwu.edu/~nsarchiv/NSAEBB/NSAEBB99/index.htm.
Economics Research Associates (1984). *Community Economic Impact of the 1984 Olympic Games in Los Angeles and Southern California* (report to the Los Angeles Olympic Organizing Committee). Los Angeles: Economics Research Associates.
Espy, Richard (1979). *The Politics of the Olympic Games*. Berkeley: University of California Press.
Flyvbjerg, Bent, Mette K. Skamris Holm and Søren L. Buhl (2002). 'Underestimating

costs in public works projects: error or lie?', *Journal of the American Planning Association* 68(3): 279–295.
Flyvbjerg, Bent, Nils Bruzelius and Werner Rothengatter (2003). *Megaprojects and Risk: an Anatomy of Ambition*. Cambridge: Cambridge University Press.
Giddens, Anthony (1991). *Modernity and Self-Identity: Self and Society in the Late Modern Age*. Cambridge: Polity Press.
Giddens, Anthony (1999). 'Risk and responsibility', *Modern Law Review* 62(1): 1–10.
Groussard, Serge (1975). *The Blood of Israel: the Massacre of the Israeli Athletes, the Olympics, 1972*. New York: Morrow.
Guttmann, Allen (1984). *The Games Must Go On: Avery Bundage and the Olympic Movement*. New York: Columbia University Press.
Guttmann, Allen (1992). *The Olympics: A History of the Modern Games*. Urbana-Chicago: University of Illinois Press.
Hacking, Ian (1975). *The Emergence of Probability: a Philosophical Study of Early Ideas About Probability, Induction and Statistical Inference*. Cambridge: Cambridge University Press.
Hacking, Ian (1990). *The Taming of Chance*. Cambridge: Cambridge University Press.
Hacking, Ian (2003). 'Risk and dirt.' In Richard V. Ericson and Aaron Doyle (eds), *Risk and Morality*. Toronto: University of Toronto Press, pp. 22–47.
Higgins, David (2007) Transcript of Oral Evidence, 5 March 2007, p. 28, HC377, Committee of Public Accounts. *Preparations for the London 2012 Olympic and Paralympic Games – Risk Assessment and Management*. London: the Stationery Office Limited.
Higgs, Robert (1987). *Crisis and Leviathan: Critical Episodes in the Growth of American Government*. Oxford: Oxford University Press.
Hill, Christopher R. (1997). *Olympic Politics: Athens to Atlanta 1896–1996*. Manchester: Manchester University Press.
Hood, Christopher (1991). 'A new public management for all seasons?', *Public Administration* 69(1): 3–19.
Hood, Christopher and David K.C. Jones (eds) (1996). *Accident and Design: Contemporary Debates in Risk Management*. London: UCL Press.
Hood, Christopher, Robert Baldwin and Henry Rothstein (2001). *The Government of Risk*. Oxford: Oxford University Press.
Hood, Christopher, David K.C. Jones, Nick F. Pidgeon and Barry A. Turner (1992). 'Risk management'. In Royal Society, *Risk: Analysis, Perception, Management*. London: Royal Society, pp. 135–192.
Hood, Christopher, Colin Scott, Oliver James, George Jones and Tony Travers (1999). *Regulation Inside Government: Waste-Watchers, Quality Police and Sleaze-Busters*. Oxford: Oxford University Press.
House of Commons Library Research Paper 05/55 (2005). *The London Olympics Bill*, 14 July.
House of Commons Public Accounts Committee, HC85 (2008a). *The Budget for the London 2012 Olympic and Paralympic Games. Fourteenth Report of Session 2007–08*. London: the Stationery Office.
House of Commons Public Accounts Committee, HC890 (2008b). *Preparations for the London 2012 Olympic and Paralympic Games, Fiftieth Report of Session 2007–08*. London: the Stationery Office.
House of Commons Select Committee on Culture, Media and Sport, HC268 (2003). *A London Olympic Bid for 2012*. London: the Stationery Office.
International Olympic Committee (2004a). *2012 Candidature Procedure and Questionnaire*. Lausanne: IOC.

International Olympic Committee (2004b). *Report of the IOC Evaluation Commission for the Games of the XXX Olympiad in 2012.* Lausanne: IOC.

Iton, J. (1978). *The Economic Impact of the 1976 Olympic Games, Report to the Organising Committee of the 1976 Games.* Montreal: Office of Industrial Research, McGill University.

Jennings, Will (2008). 'London 2012: Olympic risk, risk management, and Olymponomics', *John Liner Review* 22(2): 39–45.

Jennings, Will and Martin Lodge (2010). 'Critical infrastructures, resilience and organisation of mega-projects: the Olympic Games.' In Bridget Hutter (ed.), *Anticipating Risks and Organising Risk Regulation.* Cambridge: Cambridge University Press.

Jennings, Will and Martin Lodge (forthcoming). 'The Olympic Games: coping with risks and crises at a mega-event.' Uriel Rosenthal, Brian Jacobs, Louise K. Comfort and Ira Helsloot (eds), *Mega-Crises.* Springfield: Charles C. Thomas.

Kasperson, Roger E., O. Renn, P. Slovic, H. Brown, J. Emel, R. Goble, J.X. Kasperson and S. Ratick (1988). 'The social amplification of risk: a conceptual framework', *Risk Analysis* 8(2): 177–187.

Klinenberg, Eric (2002). *Heat Wave: a Social Autopsy of Disaster in Chicago.* Chicago: University of Chicago Press.

LA84 Foundation (2004). 'Los Angeles and the 1984 Olympic Games', www.la84foundation.org/20thAnniversary.pdf.

Lagadec, Patrick (2004). 'Understanding the French 2003 heat wave experience: beyond the heat, a multi-layered challenge', *Journal of Contingencies and Crisis Management* 12: 160–169.

Lagadec, Patrick (2007). 'Over the edge of the world', *Crisis Response Journal* 3(4): 48–49.

Lagadec, Patrick and Erwann O. Michel-Kerjan (2005). 'A new era calls for a new model', *International Herald Tribune*, 1 November.

Lenckus, Dave (2008). 'With cover in place, games set to begin', *Business Insurance* 28 July.

London 2012 bid (2004). *Candidate File.*

London Organising Committee of the Olympic Games and Paralympic Games Limited (2008). *Annual Report 2007–2008.* London: LOCOG.

Luckes, David (1997). *London Olympic Bid Feasibility Study: A Report for the NOC Meeting of May 28th, 1997.*

Luckes, David (1998). 'Appraisal of successful bids: the winning cities for 1996, 2000, 2004', 16 November.

MacAloon, John J. (1981). *This Great Symbol: Pierre de Coubertin and the Origins of the Modern Olympic Games.* Chicago: University of Chicago Press.

Magnay, J. (2005). 'Winning smiles turn to terror tears', *The Melbourne Age*, 8 July, www.theage.com.au/news/world/winning-smiles-turn-to-terror-tears/2005/07/07/1120704502059.html.

Majone, Giandomenico (1994). 'The rise of the regulatory state in Europe', *West European Politics* 17(3, July): 77–101.

Majone, Giandomenico (ed.) (1996). *Regulating Europe.* London: Routledge.

Mayor of London (2004). *London Assembly Questions on London Bid for 2012 Olympic Games and Paralympic Games – 13.10.04. Answers to Non-Oral Questions, Prepared by London 2012, GLA, LDA, TfL 18.10.04*, Q. 357/2004.

Moran, Michael (2003). *The British Regulatory State: High Modernism and Hyper Innovation.* Oxford: Oxford University Press.

National Audit Office (2007). *The Budget for the London 2012 Olympic and Paralympic Games*. London: the Stationery Office.

New South Wales, Audit Office (1999). *The Sydney 2000 Olympic and Paralympic Games: Review of Estimates*. Sydney: New South Wales Audit Office.

New South Wales Parliament (2000). *General Purpose Standing Committee No. 1*; Report No. 11, Olympic Budgeting, Paper 430. Sydney: New South Wales Parliament.

OECD (2003). *Emerging Risks in the 21st Century, An Agenda for Action*. Paris: OECD.

O'Malley, Pat (2004). *Risk, Uncertainty and Government*. London: Cavendish Press/Glasshouse.

Organisationskomitee für die XI. Olympiade Berlin 1936 e. V. (1936). *Official Report: Volume I*. Berlin: Wilhelm Limpert.

Payne, Michael (2006). *Olympic Turnaround: How the Olympic Games Stepped Back from the Brink of Extinction to Become the World's Best Known Brand*. Westport: Praeger.

Perrow, Charles (1984). *Normal Accidents: Living with High-Risk Technologies*. New York: Basic Books.

Perrow, Charles (1999). *Normal Accidents: Living with High-Risk Technologies*. Princeton: Princeton University Press, Second Edition.

Pidgeon, N.F., R.K. Kasperson and P. Slovic (eds) (2003). *The Social Amplification of Risk*. Cambridge: Cambridge University Press.

Pollitt, Christopher (1995). Justification by works or by faith? Evaluating the new public management. *Evaluation* 1(2): 133–154.

Porter, Theodore M. (1995). *Trust in Numbers: the Pursuit of Objectivity in Science and Public Life*. Princeton: Princeton University Press

Pound, Richard W. (2004). *Inside the Olympics*. Toronto: John Wiley & Sons.

Pound, Richard W. (2009). 'Risk and the Olympic Games', *University of Ottawa Seminar*, 25 March.

Power, Michael (1997). *The Audit Society*. Oxford: Oxford University Press.

Power, Michael (2004). *The Risk Management of Everything*. London: Demos.

Power, Michael (2007). *Organized Uncertainty: Organizing a World of Risk Management*. Oxford: Oxford University Press.

Preuss, Holger (2000). *Economics of the Olympic Games: Hosting the Games 1972–2000*. Sydney: Walla Walla Press.

Preuss, Holger (2004). *The Economics of Staging the Olympics: a Comparison of the Games 1972–2008*. Cheltenham: Edward Elgar.

PricewaterhouseCoopers (2002). *Business and Economic Benefits of the Sydney 2000 Olympics: a Collation of Evidence*. Sydney: NSW Department of State and Regional Development.

Rigg, Nancy J. (2002). 'Gold medal performance: safety and security at the 2002 Winter Olympics', *9–1–1 Magazine* (July/August), www.9-1-1magazine.com/index2.php?option=com_content&do_pdf=1&id=30.

Reeve, Simon (2000). *One Day in September: the Full Story of the 1972 Munich Olympic Massacre and Israeli Revenge Operation 'Wrath of God'*. New York: Arcade.

Roberts, Dexter (2008). 'Olympics security is no game', *BusinessWeek* 7 August.

Sack, Kevin (1996). 'Atlanta: Day 6; Atlanta bristles at all the criticism', *New York Times*, 25 July, Section B, p.19.

Schreiber, Manfred (1972). *After Action Report of Terrorist Activities, 20th Olympic Games, Munich, West Germany, September 1972*.

Scott, James C. (1998). *Seeing Like a State: How Certain Schemes to Improve the Human Condition Have Failed*. New Haven: Yale University Press.

Sullivan, J.E. (ed.) (1905). *Spalding's Official Athletic Almanac for 1905*. New York: American Sports Publishing Co.

Suskind, Ron (2006). *The One Per Cent Doctrine: Deep Inside America's Pursuit of its Enemies Since 9/11*. New York: Simon & Schuster.

UK Select Committee on Culture, Media and Sport, HC268 (2003). *A London Olympic Bid for 2012*. London: Stationery Office.

US–Canada Power System Outage Task Force (2004). *Final Report on the August 14, 2003 Blackout in the United States and Canada: Causes and Recommendations*, April.

Vaughan, Diane (1997). *The Challenger Launch Decision: Risky Technology, Culture, and Deviance at NASA*. Chicago: Chicago University Press.

Young, David C. (1996). *The Modern Olympics: a Struggle for Revival*. Baltimore: John Hopkins University Press.

9 The role of the private security industry

David Evans

Introduction

This chapter will outline the role that the private security industry has to play in securing major events. The London Games will be the largest private security event there has ever been in the UK, yet the industry will be faced with some serious challenges – such as adequate training, the provision of sufficient numbers and the important need to generate mechanisms for more effective partnership with public bodies such as the police.

The Private Security Industries (PSI) of many countries often provide a major contribution to that country's security through the supply of equipment both to the police and public service and to the commercial sector. Governments have come to realise in the fight against terrorism and serious organised crime that, because of limited budgets, they cannot 'do it all'. Those operating in the private sector have long known this as they provide security to many public spaces. A good example of this is in retail parks and shopping malls, both of which rely upon private sector provision.

The private security sector in any nation is diverse in the services and products it supplies to that country's public and commercial interests, which will range from guarding to security systems, consultancy to cash movement. The role of traditional private security has widened to include homeland security and resilience, and areas formerly in the defence or public arena such as CBRN protection. Prisons, policing and border and immigration control are also increasingly handled by the PSI. For example, G4S manage safe and secure short-term holding facilities at ports around the country and staff also work at Immigration Reporting Centres. Reliance Security Group operate 39 police custody suites and manage 350 custody areas within the UK court system. A total of 11 per cent of the UK's prison population are housed in prisons owned and managed by the private sector.

Security is one of the world's major growth industries, yet it is one built around steady and predicted demand. The industry is conservative in its outlook and, as such, is risk averse. Major events within developed nations are usually of a scale that lies within the capacity of the country's industry to supply. Most of the developed nations have embraced regulation of the private security industry with regard to the background of employees and companies working in it, and particularly those people guarding persons or property. This usually takes the form of licensing the individual, employer or both. Within the UK, the regulator is the Security Industry Authority whose main role is to licence individuals in areas such as Door Supervision, Security Guarding and Close Protection. They operate a voluntary approval scheme for employers. The security of major events is a mixture of security in the form of access control and searching and safety through stewarding. Stewarding has become increasingly sophisticated and the term 'Crowd Management' is often used to reflect this change

This chapter outlines and examines the role of the private security industry in providing security for the London 2012 Olympic and Paralympic Games. In doing so, the chapter is organised into four substantive areas of discussion. First, the key public players charged with supplying Olympic security are mapped. These range from state-level government agencies to more localised organisations such as the police. The second section provides an anatomy of private security provision for the Olympics. Rather than being considered as a cohesive body, the industry is comprised of a myriad of bodies holding varying degrees of scale, capacity and expertise. This section examines these components of the wider industry in more detail before considering the role of supra-industry bodies, such as Trade Associations, and the implications of regulation. The third area of discussion examines the varying requirements for security provision in the lead-up and actual staging periods of the Games. The chapter then draws out and critiques some of the important analytical themes related to private security and the Olympics. Key here are issues of cohesion, both across private sector suppliers and also with public sector agencies. Also identified is the need to transmit and build upon lessons learnt at previous Olympiads and the need for all partners to recognise the boundaries of the private security industry's capacity. The chapter will conclude with a brief restatement of the key themes and a reflection on post-2012 considerations for private security providers.

Key Olympic security partners

Security policy for the organisers of the Olympic Games is dictated by the IOC, and that policy is reflected in the host city's bid document and undergoes further scrutiny during the preparations to host the Games. Bids from prospective cities contain a specific security section outlining how they will ensure the Games take place in a safe and secure environment. The host city's national government will be required to sign off the security undertaking, and in London 2012 it is the UK's Home Secretary who will be responsible for ensuring a 'safe and secure Games' in line with the proposal that secured the Games for London. An acknowledgment of the private sector's role in the provision of security is normally contained within the bid document and, in London's case, it stated: 'the private security companies ... will have an important role in public safety and security operation' (cited in SourceSecurity, 2005).

The IOC, whilst responsible for ensuring that the host city's country is committed to the delivery of its accepted level of security, does not dictate that country's overall security policy or threat assessment. This is an important aspect as a country's national security resources are required, even though they may not have been consulted in-depth during the preparation of the bid. The result is that there can be initial confusion over responsibilities, concern over budgets and a general lack of cohesion, the latter being an issue that has generated substantial difficulties within previous Olympic security operations, such as those at Atlanta and Athens (see Fussey, this volume).

The OCOG

The organising committee for an Olympic Games is known as an 'OCOG', with the host normally denoted by the first letter placed in front (e.g. BOCOG: Beijing; LOCOG: London). The host city's OCOG is responsible for delivering its part in the bid's security plan and this is usually the element of security contained 'within venue', which includes a venue's overlay. The major part of this element will be for 'Mag & Bag' search and stewarding. It is important to understand that the OCOG is a private company – not a public body – and is charged with putting the Games on. Security, whilst important to them, in both physical and reputational terms, is understandably not the main driver.

The OCOG's budget will have been predicated in their bid and, whilst great care will have been taken to create the budget, its various elements will contain many estimates and its speed of assembly means that not all of these will have been cleared by the authorities ultimately responsible for the overall security plan. The costs will have been determined six years out and can only take a 'best guess' at inflation and the security threat assessment at the time. There is no example of any Games coming in at the original cost estimate, a feature that is particularly germane to the initial costings for safety and security.

The use of volunteers at Games time is part of Olympic culture but each bid may differ in its use of them in a security capacity and, whilst an OCOG would prefer to utilise a country's police force to carry out security roles, the cost and capacity of the police precludes this. Some bids may budget for the use of professional security guards and stewards, and others may decide to use only volunteers, whilst others may have a mix of professionals and volunteers.

OCOGs, as well as utilising new-builds, also use a mixture of existing venues ranging from exhibition halls to sports centres. These range from purpose-built to temporary to existing venues. It is quite common for organising committees to take over existing venues as a vacant premises, i.e. they provide all the services to the premises such as hall management, catering and security. Whilst it can be argued that a vacant hall and kitchens are much the same anywhere, the knowledge and experience of an incumbent security team is very difficult to replicate in the time available before a venue is taken over for the Games.

The delivery authority

Each of the Games utilises public and private money to fund new stadia. A delivery authority is usually created to coordinate activity and design, and be responsible for management through to legacy. Delivery authorities often employ an experienced private sector delivery partner to ensure that the programme is built on time and to budget. The British–American consortium CLM is managing London's build. This consortium's members have been involved in five previous Olympic Games.

Building security into the design of new stadia is fundamental to the overall security plan and the delivery authority will seek guidance from government bodies, the police and industry. Modern design techniques such as blast modelling can be (and are) used to influence the design to ensure that buildings are as safe and secure as possible (see Coaffee, this volume). The difficulty in ensuring that the design of the buildings is to the highest security standard usually comes from:

- the tight timescale in which buildings are to be designed and built;
- budgetary constraints;
- lack of early knowledge of the OCOG's security requirements;
- conflict between design and security.

The delivery authority will usually have its own security team dedicated to the project and they will be supported by security consultants. They will be responsible not only for inputting into the design of buildings but for the security of the build itself. This responsibility for the security of the build is also shared with the lead contractor(s). The speed of the design and build is usually in advance of the security planning, because of the tight timescale, and it is not unusual for the security team to be playing catch-up from the start. This will affect many elements of the plan and particularly the Olympic site where the delivery authority will find it difficult to impose a single security regime. Lead contractors are often chosen before security plans are set, and this means that those contractors have already made a financial provision for the security of the build. They will have made this provision on the basis of their experience and it can be at odds with the plans of the delivery authority. This means that either the authority accepts the level of security the lead contractor(s) is offering or demands a different level and compensates the contractor for it. This can often lead to a two-tier system of security on-site with the contractors providing their own security within its area of the site and the delivery authority providing the full site perimeter security.

The delivery authority will usually have control of security in the common areas of the site and the perimeter, and this will allow it to control access and egress to the site for workers and goods. This is of particular importance in ensuring that only authorised staff and materials enter the site. These sites are often the largest building projects in their country with a fixed timetable to completion. This will mean unprecedented movement of staff and materials, all of which will require careful planning to ensure the efficiency of the logistical part of the build. The delivery authority will have to implement search regimes that are effective (but not show-stoppers), and design a system for delivery of materials that meets the capacity of the access system. Screening of materials may take place on entry to the site or at specially designated logistic centres. To ease the flow of materials to site, the delivery authority may develop a system of vendor accreditation whereby known suppliers can confirm the security of materials and the supply line.

The security directorate

It is normal at major events for security outside venues to be the responsibility of a security directorate. This multi-agency directorate normally consists of representatives from the government, police (both national and local), other blue-light services (fire, ambulance, coastguard), transport providers, hospitals, local authorities and the utilities. It is not normal practice for the private security sector to be represented on an Olympic security directorate, an issue that may lead to tensions over the final delivery of the security strategy.

The police

The police are normally the drivers in ensuring that the Concept of Operations (CONOPS) for the event is fit for purpose, coordinated and delivered at Games time. Their planning for a Games is usually intense and extremely detailed, and there is an early recognition of the size and scale of the task which requires a significant step-up in what would be their normal planning and resourcing for large events. Working with the private sector is not part of normal practice for the police, although they may have good relationships with the private sector at events such as football matches. Planning to work with the private sector as a partner is difficult for them, as they are uncertain as to its reliability and standards. Police resources at Games time are stretched even with the inclusion of help from other police forces. The police will prioritise the demands on their resources, recognising that the public will expect 'business as usual', and advise those whom they cannot satisfy to seek alternatives.

Government departments

The Olympic Games are of such a size and complexity that governments need to be at the centre of planning. The risk to a country's reputation of failing to provide a successful and impressive Games is immense and can lead to lasting damage to the way the world perceives it. All government departments are drawn into the security planning, including sport, culture, security, transport, trade, overseas relationships and education (see Weston, this volume). Budgets become an issue as some departments are expected to provide support without extra resourcing. Traditionally, governments have avoided involving the private sector in any area of planning, preferring to rely on procurement as their method of engagement. This is because governments rarely have a good working relationship with the private sector and, because of this, they overestimate the private sector's capability and capacity. Moreover, by keeping them at arm's length, the procurement process commences too late to achieve the most cost-effective result.

The private security industry

Composition

The private security industry of any country is a diverse one, ranging from the supply of manpower to systems, security consultancy to fences. The size of company will vary from global companies to the very smallest SME. Some will be publicly quoted companies but most will be privately owned and, because of this diversity, governments have difficulty in communicating with all but the largest companies. This can mean that the most innovative ideas have difficulty in finding the light of day. In the UK, the government has recognised this in its counter-terrorism technology strategy and, through its accompanying 'Ideas and Innovation' publication, lays down its area of interest and actively encourages academia and industry to come forward with unique solutions.

Security companies generally prefer to work in a predictable area of demand and with long-term contracts. Events by their nature are short in length and occur in a way that is either feast or famine. A special and more skilled management is required to manage security and crowds at events, and so companies specialising in events only are common and, in security companies who provide event services, this is usually undertaken by specialist departments. In addition to these suppliers of stewards and security guards can be added other specialist services which range from traffic marshalling to floor-management to modelling of crowd movement.

Prime contractors and SMEs

The Olympic Games are very special because of their size and prestige. Their sheer scale poses a major challenge to even the largest of the prime contractors, and so companies will be wary of the nature and size of contracts and the impact those contracts could have on their normal business and reputation. The Games do, however, pose a tremendous business opportunity for companies in terms of profit, brand exposure, experience, new partners and exposure to a new market.

No single company can provide the overall private security services for every aspect of a major event. This is because the requirement for security will range from the requirement of the Games themselves to the additional requirements of the police, transport system, hospitals, hotels, sponsors, VIPs and many other affected organisations, companies and agencies. The organising committee will, however, appoint a prime contractor to be responsible for the security of venues.

This contractor will either be a facilities management company (for whom security is one part of their operations which they may subcontract), or will be a security company itself. This appointment is an extremely important one. Sponsorship may affect the route the organising committee takes, but it is essential that the chosen company has the resources, experience, capacity and appetite for the endeavour, which carries such a risk to the reputations of those involved. Timing of the appointment is also key because the prime contractor will require a significant amount of time to plan, prepare and resource for the Games.

There is a real anxiousness to be involved and prevent a competitor from taking advantage of the Games. This manifests itself in questions of 'How do I get involved?' and 'What is the procurement route?' In security, many SMEs wonder how they can get their technological solutions to the fore. The government and authorities are also anxious to demonstrate that they want everyone to benefit from the Games and that procurement is open and fair. New bodies may be formed to promote this, and governments may hold seminars and fairs to encourage SMEs in particular to showcase their wares. London 2012 has been instrumental in, and supportive of, CompeteFor which is a Web-based portal designed to advertise opportunities and bring companies together to bid, where appropriate, as partners.

The sponsor effect

To industry, the agreements between sponsors and the IOC or OCOG are often one of the most perplexing elements of the Games. In security terms, this can range from technology to services, and companies are often worried that they will be ruled out of involvement in the Games because a sponsor has the right of supply. These arrangements, however, normally depend very much upon the terms the sponsor has negotiated, and quite often only includes a particular element, not a whole range of goods and services. The role of the sponsor is for the Games and their venues only, so contracts for the build and all other security requirements away from the venues are often unaffected by sponsorship arrangements. Sponsors invest a considerable amount in the Games and seek to maximise their return. This means that their presence at the Games will be significant, and that they will have their own security requirement ranging from communications systems, to security systems, to close protection, to guarding that may both complement and conflict with already-complex security arrangements.

At the same time, sponsors of the Games pay a considerable sum for the privilege. They intend to make the most of the Games in order to promote their brand. As a result, they will hold numerous events, some within the Olympic Park and many within the host city and its environment (for example, VISA took more than 6,000 overseas guests to the Beijing Games). Many of their guests will be VIPs from around the world. They will have a significant demand for security, from close protection to increased security at their events and accommodation.

Role of the Trade Associations

Other key players in this equation are Trade Associations, who exist to promote the interests of their members and to provide a unified voice for their industry. These may be considered as particularly important as the view is taken that success at the Olympic Games is dependent upon the support of industry in a broader sense rather than individual companies. In security terms, the OCOG relies upon the private sector to provide security systems, people and expertise, and the government relies upon the private sector to be part of its overall security plan. Governments in their submission to the IOC on security will make mention of the private sector's role and might even call them a partner in the plan. The difficulty of this assumption is that it is rare for the public and private sectors to be partners in any venture that is short term in nature. It is a desired position easily voiced, but, without a mechanism to ensure it happens, it is unlikely to be achieved. Moreover, with regard to previous Olympiads, public sector agencies have harboured significant discomfort in working closely with the private sector. Amongst other difficulties is the fear that in collaborating on such a scale, they lay themselves open to criticism with regards to influence and corruption. As a result, partnerships are often reduced to the post-procurement process only. Private sector companies are also wary about being too involved prior to procurement, otherwise they may be barred from the procurement process altogether to prevent accusations of them gaining unfair advantage over competitors.

In order to navigate these tensions, Trade Associations are often viewed as holding a unique position to forge partnerships with the OCOGs and government departments. This is because they are one step removed from the commercial arena and are able to offer an unbiased industry view, ultimately providing a bridge between the authorities and industry without fear of complicity. The OCOGs and government can invite the Trade Associations to be part of the planning or review process. It can solicit industry feedback on the various security strands by utilising the Trade Associations to run workgroups to assess requirement, carry out analysis of any shortfall, report on capability, provide solutions and advise on the procurement timeline and process. London 2012 is witnessing unprecedented involvement of industry in the planning and review process with its Olympic Security Directorate, having an Industry Advisory Group co-chaired by the Directorate and the Resilience Suppliers' Community (RISC), the industry coordinating body.

Moreover, to promote cohesion across the industry for London 2012, the British Security Industry Association hold regular briefings for its members and has an area of its website dedicated to the Games. It contains commercially useful information to keep its members abreast of developments and opportunities.

Security industry regulation

Regulation of the security sector within a country, especially one holding the Olympic Games, is relatively common. Regulation is usually set to cover the long-term provision of security in the commercial and public sectors, not short-term events. The short-term nature of major events coupled with a massive demand for workers in stewarding, security, catering, cleaning and logistics (estimated at over 100,000 people for London 2012) puts security and event stewarding companies in a difficult position as they not only have to recruit and train staff but possibly licence them as well, and this is required to be completed prior to the start of the Games. This can create significant problems for the regulator who will have to plan for a significant increase in demand coupled with a short turnaround required to meet industry's and the Games' needs.

Private security and the Olympics: from planning to hosting

Pre-Games: requirements during the build

The delivery authority responsible for building the venues (the ODA in the case of London 2012) will lay down the security measures to be adopted during the build (see Coaffee, this volume) and they will have overall responsibility for the Olympic Park. Conflict will come, however, in the letting of construction contracts to prime contractors – an Olympic Park could have a dozen or so of these in position at any one time. Prime contractors are used to providing their own security measures and managing their own risk during any build, whereas many of the built-in security measures for the Olympic Park are mandated by the public sector agencies governing the project. This can lead to disagreement between the Authority's own security team and the contractors' teams as the main aim of the contractors is to build on time and within budget, and any delay is costly. Any such delays attributable to the Authority will lead to claims for recompense.

The delivery authority's security team, although focused on the build, will attempt during their planning to second-guess the OCOG's requirement at Games time. This will be to try to ensure that the security infrastructure (such as communications, command and control, physical measures and so on) is capable of being used or adapted by the OCOG at Games time. The reason for second-guessing is that the OCOG's security team will not begin looking at those requirements until a much later date. This is because the OCOG is financially constrained and does not invest early in the security modelling necessary to inform the delivery authority.

Requirement at Games time

Games requirements

Security at Games time in venues is determined (and paid for) by the OCOG and is scrutinised by the IOC, the host country's security directorate, and the security advisors of some participating countries such as the USA and Israel. Security measures will include the following.

- *Perimeter security of venue and overlay.* This will include the use of physical-barrier protection (fences, hostile vehicle mitigation, obstacles), electronic systems (access control, intruder detection, CCTV and Automatic Number Plate Recognition (ANPR)), pass, ticketing and accreditation systems, and security guarding. A significant element of the guarding is provided by searches. This ranges from the search of vehicles and pedestrians entering the areas of venue overlay, specific areas such as the International Broadcasting Centre, Press Centre and Olympic Village, to the search of spectators entering venues. Spectator searching is often referred to as 'Mag & Bag'. This stands for 'magnetic', which refers to the use of magnetic arch detectors or wands, and 'bag', which refers to bag rummage searches and/or the use of specialist X-ray detection equipment. These searches are not only looking for illegal items but also for the possession of items that are prohibited by sponsorship arrangements.
- *Command and control.* At Games time, this is multi-layered. The OCOG will have a central command and control centre linked to individual centres located in each venue. The police will have a separate command and control centre to coordinate both the Games requirement and the host city and country's requirement, together with links to transport command and control centres. Fire and rescue services, ambulance, health and local authorities will all have their own command centres and government too will have an overall command centre which, in an emergency, will bring the full response of the state into play.
- *Cyber security.* This is one of the most important and unseen elements of the modern Games. All communications within and without the Games are commonly IP-based. This means that it can be attacked by anyone from anywhere and, most certainly, attempts to hack into the Games systems in order to disrupt, delay or corrupt them will be made. The Games IT suppliers are well aware of the threat that is posed and a significant effort is made to ensure that the best possible virus and firewall systems are deployed. The Games are unique in being of a pre-determined length, with a high IT requirement and a reliance on volunteers to manage the system, with upwards of 30,000 users involved (see Fussey, this volume, for related difficulties at Athens).
- *Close protection.* The Games attract not only the 'Olympic family' from the competing nations, but also heads of state and presidents and also the CEOs and families of sponsors and international companies. Most of these VIPs

will be looking to have some form of close protection whilst visiting the Games or other events within the host country. Ideally, they would seek such protection from specialist units within the host country's police force. This demand is often beyond the capacity of any nation's police force and the police are normally forced to ration such protection to those under the greatest threat or for those for whom it is a Games or political requirement. This leaves those for whom police protection cannot be provided to bring their own close-protection officers or to hire them in the host country. This can be very difficult to coordinate, leading to misunderstandings, wrong assumptions and an overestimation of the private sector's ability to respond. The police will have control of the prioritisation of VIPs for close protection and they will match the list to their ability to supply. Beyond that, their advice to use the private sector may be based on little more than referral rather than an understanding of the private sector's capacity to provide.

- *Transport security.* This is one of the most significant and complex of all the security requirements of the Games. Games-specific requirements include those outlined in Table 9.1:

  The host nation's public transport system will see improved security and an enhancement will normally be obtained, or at least given substantial support, from the private sector for the period of the Games. The Games' organisers will supplement the public transport system by utilising their own hired provision of buses, taxis and chauffeured vehicles. The security of these buses and chauffeured vehicles needs careful consideration, including the period when the vehicles are empty and parked. Moreover, the background checking of the drivers and staff is a very important aspect of transport security.

- *Non-Games venues.* These will include hotels, airports, the athlete's village or accommodation, and international press and media centres. All of these venues pose particular and differing security requirements, and they will rely on the private sector to provide specific or enhanced security. These security costs will be met either by the organising committee or by the commercial supplier, for example in the hotels and transport hubs.

Table 9.1

Users	Methods
Spectators	Rail
Olympic family	Underground
Athletes	Road, including the Olympic road network
Officials	Bus
Support staff	Taxis
Suppliers	Chauffeured vehicles
Journalists	Air
	Sea and river
	Pedestrian

Non-Games requirements

As large as the Games are (they can attract more than nine million ticket holders), the events that surround the Games, and for which the Games are the catalyst, are of equal size in terms of drawing visitors.

- *Other events.* Event organisers see this number of spectators as a market to be drawn on and, as a result, they will organise events, some of which will fit into the four year Cultural Olympiad. Museums, for example, can come together to host cultural events for London 2012 (the nation's museums, libraries and archives are working together to hold events as part of the Cultural Olympiad). Individual commercial organisers will seize the opportunity to hold non-Olympic sports events (e.g. motor racing), exhibitions, conferences, festivals, rallies and shows. All of these require security and professional crowd management. The difficulty the authorities have in planning for other events is the certainty that they will happen, but the uncertainty as to what events will take place, their size and locations. Security planners have to rely on communication with local authorities to have notice of licence applications. In addition to often being excluded from early planning processes, key difficulties for private security suppliers relate to the uncertainty of the level of demand (particularly when coupled with the scale of demand of the Games themselves), that ultimately test the security industry capacity and ability to supply. A corollary of this can be many events failing to have the right level of security.
- *Additional security.* Many organisations, such as ports and airports, will seek to supplement their existing arrangements at Games time. The country's critical national infrastructure will undergo a risk assessment for the impact of a failure in their security at Games time. This will lead, undoubtedly, to a strengthening of the infrastructure in terms of systems and manpower.

Private security and the Olympics: critical reflections

Previous experience at major events

The world's major events are generally international and often sporting in nature, and will move from country to country as nations bid to host them. Security is a fundamental and important part of a country's bid. Specialist bidding companies retained to assist countries in their bids know this and seek to ensure that bid documents reflect the experience learnt from previous Games. They know that the security element of a bid will be scrutinised by the awarding body's security advisers.

The major events world is a niche business, and organising staff will move from one event to another. They tend to know (and wish to work with) those with whom they have a personal relationship and who they know have experience and can deliver. There is a general air of openness and wishing to share past experiences and best practice. Exchange visits by staff from previous events, or to events in the planning stage, or to those that are being staged, are common means of transmitting knowledge. However, building the lessons learnt into security plans is more difficult – budgets are set early and the awarding body's organisers' security procedures are more general in nature and not set necessarily to suit a particular country's own threat assessment, which will cover the wider environment beyond the venues. The awarding body audits progress in the planning stage and this will generally be according to its own security guidance, which can lead to security plans having to serve 'two masters'. As alluded to above, a consistent theme running through major events are erroneous assumptions as to the private sector's capacity to supply and, indeed, its appetite to supply – a balance often struck between marketing gain and risk to reputation. The most common error is related to the private sector's ability to supply the short-term requirement of trained stewards and security guards. The Games' needs will be in excess of 7,000 such people, yet the industry does not have this number available in its normal capacity. Indeed, most of the Games held in the past 20 years have struggled in this area. Suppliers to previous Games are often unhappy to consider involvement in future Games because of the difficulties they have encountered in supplying such a high number of staff, an issue normally outside their own or anyone else's experience.

Understanding industry capacity

Perhaps the best example of a large demand is the requirement for manpower to carry out 'mag & bag' checks and stewarding at venues. To supply the considerable number of trained (and perhaps licensed) staff for just a few days, against a background of huge and competing demand for labour, is beyond most security contractors, even if they have the appetite or experience for the role. Organisers in general do not recognise this and lessons are not learnt as Games move from country to country, but are repeated as each country overestimates the ability of the private sector to supply. This over-assumption is further complicated by sponsorship issues and a general practice of late procurement, such as those affecting the 2004 Games in Athens and 2006 Winter Games in Turin (see Fussey *et al*., 2010).

Past experience has shown that in some events manning levels have been as low as 40 per cent of requirement, and this has compromised not only security, but the event itself, by having insufficient trained stewards to manage spectators. This has often led to greater use of untrained volunteers or a call on a country's police force to cover the requirement. This can be not only extremely expensive but can also put significant pressure on the police's own resources and priorities, all of which would have been allocated to delivering the country's overall security plan.

The supply of equipment on the scale required needs sufficient notice and an understanding of industry's manufacturing process and timelines. A good example of this is in the area of technical search equipment. This equipment is expensive and requires sufficient time to manufacture and train staff to the level appropriate for its use. Organisers will want to use a minimum amount of this equipment appropriate to their own risk assessment. The host country may have a different view of the amount required, and powerful visiting countries may also have another view. These views generally express a much higher need for the equipment than that acknowledged by the organisers. Experience has shown that, together, the outcome results in a significantly higher demand than previously anticipated, and that procurement becomes a distress purchase with all the well-documented attendant difficulties of price and organisation that accompany such decisions. The lesson to be learnt from previous events on industry's capacity to supply is that industry needs to be involved and consulted in planning at an early stage in order to avoid failure or difficulties in supply.

Existing customers and 'business as usual'?

Every host country has to face the question of coordinating and drawing on its resources to meet the event's requirements and security plan. It also, however, has to recognise that its commercial community and citizens will continue to wish to carry on their normal day-to-day activities without disruption. For the commercial community, this will be a need to carry on their normal business practice without disruption, and yet they will wish to take advantage of the huge business opportunity offered through the visiting spectators and tourists. Their own demands for security will increase and they will look to their existing suppliers to meet these. The demand for additional security outside the main event is difficult to assess as it has not been studied adequately in previous major events and thus there is no blueprint to follow. It does put pressure on the supply chain to the main event as security suppliers will look to support their existing customers first. This is a natural reaction as the Games are a short event and businesses rely on the loyalty of their existing customers for their long-term success. Upsetting customers by stretching a business for a short-term gain can lead to long-term damage in terms of lost business and harm to reputation.

The existing customers of security companies, especially those supplying services to the Games and those within the financial and commercial districts of the host city, will not only demand 'business as usual' from their suppliers, but will also seek additional cover. Increased demand for security is not limited to the host city. When coupled with the high demand for security services for the Olympics, potential tensions may develop over the provision of security to both existing customers and other members of the public. Many national and international companies understand that police resources will be stretched across the whole country and, as a result, will seek to increase private provision. For example, major retailers will forecast a drop in police response to incidents and will seek to deal with the potential lack of response in other ways. Companies reliant on CCTV and intruder and access systems will fear the effect of a shortage of skilled technicians at fit-out at Games time and will seek to introduce incentive schemes for their maintenance technicians to ensure that their systems remain effective.

Private security and planning: towards a holistic approach to security?

Major events, in a world where the risk of terrorism is high and where that terrorism looks to create mayhem in crowded places or infrastructure, and seeks publicity through 'spectaculars', do require a special national security plan, in addition to the event's own security plan. This plan needs to be holistic in nature and to garner together all the different elements that make up a national infrastructure. This will range from the police to transport, to intelligence services, to the armed forces, to government and local government, to utilities, to health *and* to the private sector.

Whilst government agencies will, in many instances, have worked together, it is unusual for the private sector to be involved. And yet, the private sector may account for over 60 per cent of the provision of security at event time in areas ranging from security at the venues to sponsors, to hotels, to the transport and logistic systems, and to the Cultural Olympiad. To incorporate the private sector in the planning and preparation of major events does not come naturally to government bodies. They can have a jaundiced view of the private sector as bodies only interested in making money, and this can lead to them being treated as a procurement-only issue. To involve the private sector in the security planning of major events requires recognition of its importance and of the benefits it can bring. Even then, difficulties may arise over a range of issues, including private industry members not holding the security clearance necessary to attend certain meetings and briefings or to view classified documentation.

- *Partnerships.* Partnership is at the heart of all major Games – from the organising committee, to collaboration with the public sector, to the integration of systems and services. This is why companies with previous Games experience and existing partnership arrangements are favoured over those companies without them. For example, Atos Origin is the IOC's Worldwide Information Technology Partner and provides one of the most critical elements of the Games, and the IOC recognises the importance of that service. Atos Origin was appointed for the 2012 Games before the Games were awarded to London. All major events are time- and cost-critical with a significant risk to reputation if things go wrong. Sponsors often have the relevant experience, and their appointment has a significant effect on the procurement process. In addition, managing expectation of industry is a major challenge as all industry expects to have opportunities from the holding of a major event, when, in fact, a previous track record is one of the major criteria in the choice of supplier.

  Integration of the supply of services is key and it is often those partnerships tying together logistics and systems that dictate how and who is involved from industry. The major event industry is a relatively small industry with global reach. The players in it do know each other and the partnerships are often global ones with teams moving from one event to another. This event experience is often in specialist areas with teams allocated to the Olympics or other major sporting events.

- *Private sector cohesion.* The private sector, apart from some partnerships for specific requirements, does not generally work collaboratively with competitors. Those in the sector prefer to see competitor companies fail rather than succeed. In the period of a major event, such as the Olympic Games, most companies will seek to supply prime contractors with goods and services, and to meet the demand of its existing customers. Whilst wishing to see competitors fail, companies do wish to see their country do well and do want the major event to succeed.

  This wish can be turned into collaboration for this one occasion and the route to this is through the Trade Associations who can act as the honest broker and organiser of that collaboration (see above, p. 170). Companies in Trade Associations routinely work together on the development of standards and other areas of common interest, and the good personal relationships that develop can be utilised for working for a single purpose during Games time.
- *Integration with the public sector.* The public and private sectors suffer from a mixed relationship – with the former very comfortable with the role and innovation of product suppliers but very uncomfortable with the security service providers. This is because the public sector has concerns about the standards and consistency of service and about allowing the private sector to become involved in supply areas traditionally carried out by themselves. The public sector does, however, recognise the limits of their capacity to supply a major event and the need to work with the private sector in order to meet the overall requirement. The size of the event does lead to more collaborative working, and this is something the private sector welcomes as it sees it as an opportunity to have a better long-term relationship with the public sector as a result of the event.

Concluding points: 2012 and beyond

In sum, the experience of private security at the Olympics has revealed a number of important lessons. Principal amongst these is the importance of inclusive planning. This will develop relationships whereby industry will receive the maximum lead-in times for its planning processes and, also, those in the public sector will be able to base their decisions on a realistic assessment of the ability and capacity of the private security industry. Such inclusivity will also allow the multiple and diverse agencies involved in Olympic security planning to negotiate any differences, whilst developing a shared understanding of the issues from an early stage. All major events are for a short period only. The build-up may be over a period of several years, but the actual event is only for a few weeks. It is the legacy that is important.

For the private security sector, they see the legacy in terms of:

- new relationships and new markets;
- a closer relationship with the public sector;
- new ways of working.

Major events, and especially the Olympic Games, can change ways of working forever and can introduce new opportunities. It is the forward-looking companies that recognise this, and they seek to use the Games to drive their businesses forward.

References

Fussey, P., Coaffee, J., Armstrong, G. and Hobbs, R. (forthcoming 2010) *Sustaining and Securing the Olympic City: Reconfiguring London for 2012 and beyond.* Aldershot: Ashgate.

SourceSecurity (2005) *Thousands Needed For Olympics*, available from www.sourcesecurity.com/news/articles/160.html retrieved 4th February, 2010.

10 The challenge of inter-agency coordination

Keith Weston

> The UK Police Service has a well deserved reputation for ensuring that major sporting events pass off safely. Staging the Olympics is a tremendous honour, and the police will be playing their part in ensuring that the Games are safe and secure so that spectators and participants can really enjoy this unique event ...
> (Chris Allison, Assistant Commissioner Central Operations, Olympics and Paralympics, Metropolitan Police[1])

In an ideal world, during the run-up to an Olympic Games, there would be no need to plan or prepare for the possibility that the Games might be disrupted by terrorists, serious and organised criminal gangs, or some other malicious malefactor. The tradition of the 'Truce' or 'Ekecheiria' was established in ancient Greece in the ninth century BC by the signature of a treaty between three kings. During the Truce period, the athletes, artists and their families, as well as ordinary pilgrims, could travel in total safety to participate in or attend the Olympic Games and return afterwards to their respective countries. As the opening of the Games approached, the sacred truce was proclaimed and announced by citizens of Elis who travelled throughout Greece to pass on the message.[2]

Unfortunately, this noble tradition has not been respected by those who, for a variety of reasons, would seek to either exploit the Games for criminal advantage or undertake a murderous attack to seek publicity for their cause. For these reasons, much effort is expended by host governments and various organisations to ensure that vulnerabilities are reduced, thereby limiting opportunities for criminals and terrorists to achieve their purpose.

The intention of the UK government and the International Olympic Committee is to ensure that the 2012 Olympic and Paralympic Games are safe and secure for athletes and spectators alike. Indeed, a major factor in the decision to bring the 2012 Olympic and Paralympic Games to London was the commitment from the UK government to ensure safety and security at every Olympic location and venue. The London Host City Contract includes commitments from:

- the Prime Minister, who signed a guarantee to 'take all financial, planning and operational measures necessary to guarantee the safety and the peaceful celebration of the Games';

- the Home Secretary, who guaranteed to 'co-ordinate all matters of security and the emergency services for the Games'; and
- the Chancellor of the Exchequer, who signed a separate guarantee to provide 'all necessary financial support to the Games, to include ... bearing the costs of providing security'.[3]

London 2012's ambition is to create a Games for everyone, where everyone is invited to take part, join in and enjoy the most exciting event in the world.[4] In an uncertain world, it is essential that any events, malicious or otherwise, that could disrupt the Games are prevented, or if this is not possible, their effect is minimised.

In the aftermath of the tragic death of Jean Charles de Menezes at Stockwell Tube Station on 22 July 2005, Her Majesty's Inspectorate of Constabulary (HMIC) conducted three reviews of the Metropolitan Police that focused on the circumstances surrounding his death and whether recommendations from previous reports had been implemented. The first review, in September 2007, requested by the then Deputy Commissioner Sir Paul Stephenson, now Commissioner of the Metropolitan Police Service, remains 'confidential'. The two later reviews, in January 2008[5] and April 2009,[6] were requested by the Metropolitan Police Authority (MPA). The focus of the third report was on Command and Control and Interoperability. To meet the public's expectations of police capabilities, and address the themes that emerged from the Independent Police Complaints Commission (IPCC),[7] HMIC, MPA[8] and Coroner's[9] reports, HMIC adopted a simple working definition for 'Inter-operability' – 'a seamless working relationship between different units and personnel'. The report considered that the seamless operation, in turn, required certain features:

- shared ethos (of what matters most);
- common doctrine (a common set of operating principles or guidance);
- a unified command model (clarity of who is in charge of what);
- compatible and reliable communications systems;
- shared language (that ensures common understanding in pressurised operating environments);
- common equipment;
- common standards of professional practice.

All supported through a process of continuous capability building by:

- shared training and exercising;
- shared operational experience through working together;
- shared learning and debriefing.

It is argued that this model of interoperability should be adopted for those planning the 2012 Games. The model will be reviewed at the conclusion of this chapter.

The purpose of this chapter is to review the coordination challenges that face the different agencies preparing to ensure a safe and secure Games in 2012, especially as they are predicted to be the biggest sporting event in UK history.[10] Initially, the role of government will be considered before moving on to the following specific coordination challenges: communication; leadership; logistics and resource management; public relations; and, finally, planning, training and exercising.

The role of government

The London 2012 *Olympic and Paralympic Safety and Security Strategy* clearly defines who is responsible for what at the planning, preparation and delivery stages of the Games. Responsibility for discharging the guarantees made by the Prime Minister and Home Secretary has been delegated to the following:

- the Ministerial Committee on Economic Development Sub Committee on the Olympic and Paralympic Games (ED(OPG)), which is responsible for coordinating and overseeing all issues relating to the London 2012 Olympic and Paralympic Games;
- the Ministerial Committee on National Security, International Relations and Development Sub-Committee on Protective Security and Resilience (NSID(PSR)), which gives direction and, where appropriate, resolves issues relating specifically to the Safety and Security Strategy, monitors the delivery of the Safety and Security Programme and reports to ED(OPG) on progress;
- the Minister for the Olympics, who has direct responsibility for the delivery of the government's overall Olympic and Paralympic Programme, and is an invited member of NSID(PSR), a permanent member of ED(OPG) and reports directly to the Prime Minister;
- the Home Secretary, who is lead minister for Olympic and Paralympic safety and security, and is accountable for the delivery of a Safety and Security Strategy, Delivery Plans and the Safety and Security Programme. The Home Secretary chairs NSID(PSR) and is a member of ED(OPG). This accountability does not override existing Ministerial statutory obligations – for example, those held by the Secretary of State for Transport, who is responsible for the security of UK transport systems and is empowered by legislation to require the regulated transport industry to implement security measures designed to protect its infrastructure, hardware, staff and the public using it from attack. Nor does it supplant other regulatory responsibilities for safety, for example the Health and Safety Executive's (HSE) regulatory role in both construction and providing advice on major hazards;
- the Senior Responsible Owner (SRO) for the Strategy, who will be responsible for its development and, through other agencies, departments and organisations, for its implementation, and who will be part of the Office for Security and Counter Terrorism (OSCT) in the Home Office;
- the Olympic and Paralympic Security Directorate (OSD) within the OSCT, which will develop and manage the Strategy and its associated programmes

The challenge of inter-agency coordination 183

and ensure their delivery through other agencies, departments and organisations under the authority of the SRO;
- the Government Olympic Executive, which is responsible for overseeing the delivery of the entire Olympic and Paralympic project and driving the public sector effort. It is accountable to Parliament and reports directly to the Minister for the Olympics;
- the London Organising Committee of the Olympic and Paralympic Games (LOCOG), which is responsible for preparing and staging the 2012 Games, and has specific responsibility for the safety of spectators and the provision of routine safety and security measures as laid down in existing legislation; and
- the Olympic Delivery Authority (ODA), which is the public body responsible for developing and building the new venues and infrastructure for the Games, including designing in safety and security measures, in accordance with statutory requirements and risk-based judgements made in consultation with other security partners.[11]

The myriad of initials and acronyms appear at first glance to be an alphabet soup. Figure 10.1 is an attempt to unpick the strands and place the various committees, sub-committees, individuals and organisations into their relative positions, within four generic areas:

- policy oversight;
- security policy;

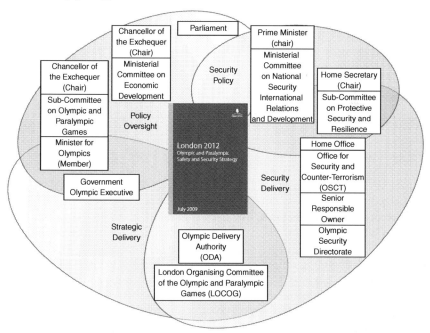

Figure 10.1

- strategic delivery; and
- security delivery.

It will be seen that these areas overlap, but for each area there is a lead Minister, with all unresolved issues being dealt with by the Prime Minister.

At the time the London 2012 *Olympic and Paralympic Safety and Security Strategy* was published, the greatest threat to the security of the Games was perceived to be international terrorism. The planning assumption was also based on a threat level of 'severe',[12] subject to future changes.[13] In the same month that the *Strategy* was published, the Joint Terrorism Analysis Centre (JTAC)[14] reduced the overall threat level to UK interests to 'substantial'.[15] The decision to plan on an assumed threat level of 'severe' is a sensible one, as it provides some stability for operational planners and decision-makers. If planners had to constantly change plans in accordance with the threat level, which are predominantly driven by intelligence, or events, they would have no consistency or stability on which to base their planning strategy. In an Olympic context, it will be important to understand exactly what moving from 'severe' to 'critical' means. It may not be a little adjustment in some areas. Nevertheless, there would be less adjustment from 'severe' to 'critical' than from 'substantial' to 'critical'.[16] It is always easier to scale down than to scale up, particularly at short notice.

To provide a framework for the successful delivery of the Olympic Safety and Security Strategy, five key Objectives have been identified, together with a Programme attached to each Objective.[17] The Objectives are:

- Protect;
- Prepare;
- Identify and Disrupt;
- Command, Control, Plan and Resource; and
- Engage.[18]

The Protect Programme is focused on protecting the venues, events and the supporting transport infrastructure, as well as people attending the events and using the facilities. It is concerned mainly with protective security measures. The Prepare Programme addresses external threats that may significantly disrupt the safety and security of the Games. It is focused on contingency planning and seeks to ensure that capabilities will be in place to mitigate the impact of a disruptive event. The Identify and Disrupt Programme will concentrate on anticipated threats to the Games. It is intelligence-focused and will rely on the existing tried-and-tested intelligence agency/law-enforcement partnership. The Command, Control, Plan and Resource Programme will be considered in more detail within this chapter. Finally, the Engage Programme is concerned with promoting the strategy with international partners and UK stakeholders to inform and advise them of the strategy, and seek their requirements and endorsement of the strategy, thereby guaranteeing support for and confidence in the safety and security arrangements that will be in place in 2012.

A key planning element underpinning the strategy will be a dynamic risk-assessment. The Olympic Safety and Security Strategic Risk Assessment (OSSSRA)[19] is intended to evolve as the recognition and comprehension of risks develops. The intention is that the risk assessment will be able to support strategic decision-makers and provide a justification for prioritising resources. OSSSRA is based on a standard Cabinet Office methodology of risk, on the basis of likelihood (threat and vulnerability) and impact.[20] This complexity of risk was discussed in 2005, following the award of the Games to London. Jennings noted that the apparent ubiquity of Olympic risk is an inevitable symptom of the increasing complexity of the physical and organisational architecture of modern Olympic Games. London 2012 is no different in this regard. Such risk carries with it political and reputational risks attached for government and organisers. Jennings concluded that, whilst there is no reason why this should not be 'the best games ever', he also believed they may be the riskiest yet.[21]

The OSSSRA is divided into four elements: a Strategic Threat Assessment; a Vulnerability Assessment; an Impact Assessment; and a Hazard Assessment. The Strategic Assessment will be owned by the Olympic Intelligence Centre (OIC). The Olympic Intelligence Centre concept was introduced at the Atlanta Olympics in 1996 and has been used in all subsequent Olympics.[22] The OIC will constantly review and update the risk assessment from four pre-defined threats: terrorism, domestic extremism, serious and organised crime, and volume crime. The Strategic Threat Assessment will draw on individual intelligence agency assessments to provide a complete picture of the risk posed by potential threats from any of the pre-defined areas. The Vulnerability Assessment will consider potential weaknesses in physical, personnel and electronic security and what measures could be taken to reduce the vulnerability. The Impact Assessment will adopt the standard Cabinet Office approach and use the headings: fatalities/casualties; disruption to essential services and infrastructure for the 2012 Games; psychological impact and reputation. Finally, the Hazard Assessment will consider the threat, vulnerability and impact of non-malicious events, such as accidents, legitimate protest and industrial action.[23]

Professor John Grieve (former Deputy Assistant Commissioner), in his foreword to *Policing Critical Incidents*,[24] lists six risk and analysis models that proved effective during his Critical Incident Management and Leadership training:

- CCA – Comparative Case Analysis;
- PLAN BI – Proportionate, Legal, Accountable, Necessary – acting on Best Information;
- EEP – Experience, Evidence, Potential;
- SAFCORM – Situation, Aims, Factors involved, Choices, Options, preferred Risk and Monitoring; the STEEPLES model below is used to assess the 'Factors involved' element of SAFCORM;
- STEEPLES – Social Technical, Environmental, Economic, Political, Legal, Ethical, Safety and health;
- RARARA – Record, Analyse, Remove, Accept, Reduce, Avoid/Averse.

Grieve noted that risk assessments were tools for avoiding critical incidents and for critical incident management when they arose. Further, he observed that recording decisions and their rationale were a form of risk, gap or threat analysis. They needed to be ethical, transparent and inclusive, based on evidence and intelligence, and wide-ranging with a foundation on existing contacts within the local community. They needed to be supported by prior arrangements for the exchange of information about concerns, and needed creativity to help explain a complex decision-making environment that contained anomalies and ambiguities, particularly where it was not always possible to supply specific information.[25] It is vital that OSSSRA incorporates the methodology developed by Grieve and others, and develops it as a dynamic risk-assessment, flexible to new threats and hazards as they evolve, but also capable of providing a stable framework to inform both contingency planners and operational decision-makers.

In January 2009, Baroness Neville-Jones asked Lord West which departments, agencies, public bodies and other organisations participated in or contributed to the Olympic safety and security programme. Lord West stated that a wide range of organisations contributed to the safety and security programme for the London 2012 Olympic and Paralympic Games, and published a list of 49 principal stakeholders, at the same time pointing out that the list was not exhaustive (see Appendix).[26]

Examination of this list highlights the many organisations that will have a role to play in ensuring the safety and security of the 2012 Games. The challenge for the Home Office will be to seek universal understanding of and compliance with the *Olympic and Paralympic Safety and Security Strategy*, and to ensure there is coordination and cooperation at every level and during every stage, from preparation to delivery of the Games. Such support and inter-agency teamwork will be vital to the detection, disruption and/or deterrence of any terrorist event intended to cause injury, alarm and distress during the Games.

A wise investment for each of the organisations mentioned in Lord West's list would be the early appointment of a senior responsible owner,[27] to be accountable for that organisation's safety and security strategy in support of the Games, and to be the single point of contact with the police and other organisations.

The Department for Culture, Media and Sport (DCMS) is the lead government department for delivering the 2012 Games. The Minister for the Olympics and Paralympics is responsible for the Olympic and Paralympic Programme and legacy plans, and co-chairs the Olympic Board (with the Mayor of London). The Board, which also consists of the chairs of the British Olympic Authority (BOA) and LOCOG, provides oversight and strategic coordination of the 2012 Games project.

The Home Office has lead responsibility for security at the Games. The development of the Olympic Security Programme is being led by the Olympic Security Directorate, located within the Home Office's Office for Security and Counter-Terrorism. The Home Office hosts the Olympic and Paralympic Security Directorate (OSD), which was formerly part of the MPS. This recognises that ownership of the strategy, and of the cross-government coordination required, needs to sit with the Home Office rather than an individual police service. The OSD will commission projects to deliver safety and security out to agencies, which will deliver those projects to an agreed budget and scope. The MPS and ACPO have joined up to create the Olympic and Paralympic Policing Co-Ordination Team (OPC), which will receive commissions on behalf of the police service. The OPC is a national team, overseen by Assistant Commissioner Allison and headed by Commander Richard Morris, responsible for coordinating the service-wide policing issues relevant to the Olympics.[28]

Having considered the top-down organisational structure created to manage the Olympics in 2012, identified the principal stakeholders and reviewed the risk-management methodology that will be used to implement the Safety and Security Strategy, it is now worth identifying the individual elements that may prove to be a challenge to inter-agency coordination. Also, it is worthwhile considering where some of these challenges have arisen in the past, and whether any lessons identified have become lessons learned. There are many examples following catastrophic events that have been the subject of post-event formal public enquiries, prosecutions or inquests, where the final report contains many well-founded recommendations that are not implemented and the identified failing is subsequently repeated. It is crucial, with the pre-planning currently taking place to ensure a safe and secure Olympic Games in 2012, that any future chair of an inquiry, judge or coroner does not have the onerous task of asking why recommendations from previous catastrophic events were not implemented.

The following areas have proved, at times, to be challenging in multi-agency responses to crises and disasters: communication; leadership; logistics and resource management; public relations; and, finally, planning, training and exercising. In 2006, two American researchers compared eight serious incidents in the United States to examine why 'emergency service organisations find it difficult to learn certain lessons – and to better understand why this is the case ...'[29] Their intention was to identify persistent challenges, across a range of different types of disasters. Table 10.1 lists the incidents they studied, together with common lessons identified, but not learned.

Donahue and Tuohy concluded that, if solutions were evident, emergency response professionals would have adopted them long ago. This should motivate agencies in all emergency response disciplines and at all levels of government to give serious attention to the goal of inculcating a culture of learning from past disasters to prevent future losses.[30] Given the planning time available for the 2012 Olympics, it is vital that contingency planners take heed of and implement recommendations from reports into previous incidents or disasters. A failure to effectively plan and prepare in a multi-agency environment would not only be unacceptable, but also unforgiveable. It will be noted from Table 10.1 that Leadership, Planning, Public Relations and Resource Management topped the list of important topics that were addressed in several of the prominent incidents they studied.[31] As previously noted, in addition to the areas identified in the US

Table 10.1 Common categories of lessons

Lessons Learned issues	Anthrax attacks	Columbia recovery	Columbine	Hurricane Katrina	Oklahoma City bombing	SARS	11 September	Sniper investigation
Communicators		X	X	X		X	X	
Leadership	X	X	X	X	X	X	X	X
Logistics	X	X		X	X	X	X	
Mental health					X		X	
Planning	X	X	X	X	X	X	X	X
Public relations	X	X	X	X	X	X	X	X
Operations		X	X	X	X	X	X	X
Resource management	X	X				X	X	X
Training and exercises	X	X	X	X	X		X	X

Note
Correlation between After Action Reports for selected major incidents and significant issues addressed.

research, the following areas have also consistently provided challenges to multi-agency responses to crises and disasters: Communication; Logistics; and Training and Exercising. All these areas will now be considered in more detail.

Communication

The National Police Improvement Agency (NPIA), in its 'Guidance on Multi-agency Interoperability', issued in 2009, found that a recurring theme from debriefs of events, incidents and exercises was the inefficiency of same-service and inter-agency communications. Conversely, on the occasions when it worked well, the participants spoke highly of the benefit of being able to exchange information rapidly in a way that was timely and useful to the parties receiving it. Organisations and their executives have been criticised by the public and the media over inefficient communications at high impact incidents and events. Public enquiries and the civil and criminal courts have held organisations to account for breaches of health and safety where ineffective communications and the absence of a common operating procedure were considered to be contributing factors. The NPIA guidance warned that if the way in which an organisation's activities are managed or structured cause a person's death and amount to a gross breach of a duty of care owed by the organisation to that person, the organisation may have committed the offence of corporate manslaughter under the Corporate Manslaughter and Corporate Homicide Act 2007.[32]

There are many examples where failures of communication have resulted in death, or have otherwise hampered the efforts of emergency service workers. It is important to understand where the failures have occurred and to ensure that they are not repeated. Alison and Crego identified six specific areas to address to improve communication and information sharing during critical incidents.[33] They include the:

- development of a common language;
- use of technology to speed communication;
- development of clear processes and protocols to assist communication and understand responsibilities to share information;
- dissemination of accurate information quickly and clearly;
- conduct of regular multi-agency meetings not only during crisis situations but also before critical incidents occur;
- commitment to genuine multi-agency partnerships, including voluntary agencies and communities.

One of the many lessons that emerged from the terrorist attacks in London on 7 July 2005 was the inability of the emergency services to communicate through a common radio system.[34] The emergency services addressed this deficiency by implementing a standard radio system, 'Airwave'. The Greater London Authority reported in August 2007 that key improvements had been made relating to communications within and between the emergency and other responding services in the initial stages following an incident.³⁵

Airwave is a digital trunked radio service for police and other emergency services in England, Scotland and Wales provided by Airwave Solutions Limited under contract to the NPIA. Airwave has replaced outdated, individually run force analogue radio systems with a national digital radio service. It is now fully established and network performance is exceeding contractual levels.³⁶

In 2005, NPIA commissioned ProLingua Ltd to deliver the AirwaveSpeak project. The aim of the project was to prepare a standard procedure and phraseology for the use of the 'Airwave' digital radio technology reflecting the opportunities and constraints of the technology and providing for the common operational demands of the police forces of the United Kingdom. The procedure and phraseology would also take into account the operational demands of other agencies deploying the technology in joint operations with the police. AirwaveSpeak began its role out to the police service in 2007; it offers a number of benefits:

- a clear radio language using everyday terminology;
- clear and unambiguous transmissions when different organisations are interoperating at major and national events;
- a potential cost–benefit as verbal clarity in initial transmissions and shorter transmissions equate to fewer repeated messages and less 'air time' being used, thus reducing costs;
- standardisation of terminology should result in improved radio discipline and assist communications room staff in terms of recording clear information;
- improved opportunity for users by reducing unnecessary lengthy voice transmissions.³⁷

On 27 January 2009, the London Organising Committee of the Olympic Games and Paralympic Games confirmed that Airwave would provide private mobile radio services to all Games venues. In preparation for the 2012 Games, Airwave is building a completely new and independent communications infrastructure that will be used from 2011 right up to and during the period of the Games. Like its public safety infrastructure, the service will be delivered using Terrestrial Trunked Radio (TETRA)³⁸ technology, but it will be totally separate. The new service will provide coverage allowing the 2012 Games and its workforce to operate effectively and safely. The infrastructure will predominantly provide coverage for the London venues but also includes venues across the UK.³⁹

The roll-out of Airwave and the AirwaveSpeak project are encouraging developments. Nevertheless, the London Assembly 7 July Review Committee, in their follow-up report, published in August 2007, cautioned against the over-reliance on one communication system. They recorded their concern that London's emergency services were putting all their eggs in one basket by relying on Airwave radios. If Personal Role Radios and/or other similar technologies are not a viable option, then clearly they should not be used. But the Committee was not convinced that serious consideration had been given to their potential use, and this gave them cause for concern.[40]

Alison and Crego acknowledged that communication is crucial to the successful resolution of critical incidents. They identified the following problems that hindered effective and efficient multi-agency working:

- failures or delays to promptly respond to requests for information;
- failure to share information due to lack of clarity as to whether information could be shared with partner agencies;
- ambiguous, unclear or misleading information being passed to partner agencies;
- poor record-keeping of incidents, conversations and meetings;
- information overload, i.e. too much information being passed between agencies;
- failure to share information due to a lack of process, or deliberately not being shared;
- failure to share information within organisations;
- failure to share information at an early stage, or because of distance between agencies or inability to contact partner agencies.

Communication is not just about passing messages by telephone, radio, fax or email. It is about ensuring the right people have the right information at the right time, to enable multi-agency decision-makers to have a common operating picture from which they can agree a course of action that will prevent, mitigate or resolve critical incidents.

Leadership

Donahue and Tuohy identified that 'uncoordinated leadership' was a common feature of the crises and disasters they studied. They found that large incidents demanded that robust command and control structures emerged out of the initial chaos that inevitably ensued when disasters struck. Large incidents also involved a multitude of agencies, each of which must direct its own resources. As a result, agency- and/or function-specific command structures proliferated. Since each agency had legitimate missions, responsibilities and jurisdiction, each used its command and control process to take charge in a legitimate attempt to solve the problems the agency was supposed to solve. What was absent was an overarching command structure to which all participants subscribed, which resulted in duplicative and conflicting efforts. As one responder put it, 'People ask "who's in charge?" The response was usually, "Of what?"'[41]

The planning for the 2012 Games is based on tried-and-tested standard operating procedures. The safety and security operation will comprise a series of locally commanded but nationally coordinated operations. The emergency services (police forces, the fire and rescue service, and the ambulance service) will have specific responsibilities for safety and security during the Olympic and Paralympic Games consistent with existing statutory and common-law obligations.[42] Therefore, the duplication, inter-agency rivalry and competition identified in the US incidents by Donahue and Tuohy are not anticipated to be issues in 2012.

In the UK, there is a well-defined structure for the management of multi-agency major incidents, whether they are naturally occurring events, such as floods, non-malicious events, such as the explosion at Buncefield on 11 December 2005, or malicious incidents, such as the terrorist bombs in London on 7 July 2005. Although every major incident is different, they can generally be broken down into four stages; initial response, consolidation phase, recovery period, and return to normality (see Figure 10.2).

- Initial response: when the emergency services have been alerted and are on their way to the scene.
- Consolidation phase: emergency service personnel at the scene will begin to identify exactly what has happened. Police officers are trained to use the acronym SAD CHALET[44] to make a rapid assessment of the scale of the incident, including the number and seriousness of casualties; whether there are additional hazards present, e.g. spilled fuel, unexploded terrorist bombs, chemicals, etc.; the best route to the scene and location of any rendezvous points; and inform the control room of the type and extent of additional resources required. The other emergency services will undertake their core roles, including extinguishing fires, rescuing injured people and treating them before removal to hospital. The police will concentrate on isolating the scene and removing all unauthorised civilians from within the outer cordon, whilst identifying and clearing essential routes for emergency service vehicles. Depending on the seriousness of the incident, a Joint Emergency

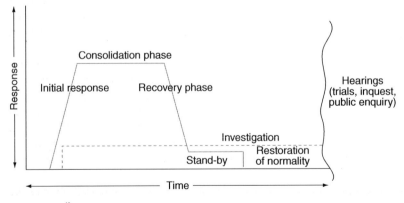

Figure 10.2[43]

Services Control Centre will be established to coordinate the response to the incident, which will include, as appropriate, local authorities and utility providers. If crime is suspected, terrorist or otherwise, a Senior Investigating Officer will be appointed to conduct an investigation from within the inner cordon.

- Recovery phase: during this phase the emergency service's mission is to bring the incident under control, either from within existing resources, through assistance from surrounding emergency services (under pre-existing mutual aid agreements) or by recalling off-duty staff.
- Return to normality: depending on the scale and severity of the incident, this may take hours, days or weeks. During this period, there will be a gradual handover of command and control from the emergency services to local authorities, health agencies, private cleaning contractors, etc.

During the various stages of a Major Incident, a Gold, Silver and Bronze command and control structure will be implemented. Gold, Silver and Bronze are titles of functions adopted by each of the emergency services and are role-related, not rank-related. These functions are equivalent to those described as Strategic, Tactical and Operational in other documents about emergency procedures.[45] In summary, the roles of each can be described as follows. Gold is the commander in overall charge of each service, responsible for formulating the strategy for the incident. Each Gold has overall command of the resources of their own organisation, but delegates tactical decisions to their respective Silver(s).

At the outset of the incident, Gold will determine the strategy and record a strategy statement. This will be monitored and subject to ongoing review. Silver will attend the scene, take charge and be responsible for formulating the tactics to be adopted by their service to achieve the strategy set by Gold. Silver does not become personally involved with activities close to the incident, but remains detached. Bronze will control and deploy the resources of their respective service within a geographical sector or specific role, and implement the tactics defined by Silver. As the incident progresses and more resources attend the rendezvous point (RVP), the level of supervision will increase in proportion. As senior managers arrive, they will be assigned functions within the Gold, Silver and Bronze structure.[46]

Kevin Arbuthnott, a retired senior UK fire service officer, has critically analysed the Gold, Silver and Bronze model. He commented that every incident is unique in some respects, and different agencies and jurisdictions will need to resolve them in their own way. Nevertheless, the concept of multi-agency incident management, and the synergies that result from effective joint operations, rely heavily on officers of all the responding agencies understanding what is being demanded of, and by, their partner organisations at the multi-agency table.[47]

Whilst every effort will be focused on ensuring that the 2012 Games are remembered as a fantastic sporting event, there is a risk that terrorists will attempt to disrupt the event. A terrorist attack during the 2012 Olympics, whether specifically targeted at an Olympic venue or not, will activate a Major Incident response from the emergency services who will adopt the procedure described above to manage and mitigate the consequence of any such attack. The emergency service personnel will be trained and experienced in a multi-agency response to Major Incidents and, during the lead-up time to the Olympics, will undergo further specialist training, planning and participation in a variety of exercise scenarios.

Logistics and resource management

The key to the management of logistics and resources for a major sporting event, such as the Olympics, is to identify what is needed, where it is needed, when it is needed, why it is needed and who needs it – and then to deliver or supply the goods or services in a safe, secure and usable manner. In a normal environment, satisfying these requirements is a routine part of supply chain management. Should terrorists decide to disrupt the Olympics through an attack on the supply chain,[48] the potential for disruption will be enormous.

The challenge for those implementing the Olympic Safety and Security Programme will be to ensure a balance in the security regime. On the one hand, it will be crucial to reduce any potential vulnerability in the supply chain and secure it from any possible terrorist attack, whilst at the same time not imposing a comprehensive security regime that is so draconian that the supply chain is disrupted by constant searches and other security checks. Early engagement with suppliers of goods and services will be necessary to ensure that a close cooperative partnership is created which will include an understanding and awareness of the security measures that will need to be implemented during the period of the Games.

Ensuring that goods and services are supplied to the Games participants and spectators in a safe and secure manner is not the only logistical and resource management challenge that needs to be addressed. The Olympic Safety and Security Strategy Command, Control, Plan and Resource Programme includes specific reference to the requirement to identify, train and otherwise prepare sufficient human resources to be available for the Games. This programme will also plan, procure and test the logistical requirements to marshal and deploy those resources and define how they will be managed, at all Olympic and Paralympic venues and supporting sites.[49] The intention is to establish a series of secure strategic multi-agency logistic support centres to ensure that the emergency services have sufficient support available to meet the requirements of any critical incident.[50]

Police personnel provided under the existing mutual aid arrangements[51] will be coordinated by the Police National Information and Coordination Centre (PNICC). In 2003, the Association of Chief Police Officers (ACPO) established PNICC to coordinate the national mobilisation of police resources. Funding for PNICC is provided via contributions from police authorities rather than directly from the Home Office. This funding provides for three permanent PNICC staff and associated costs. Additional staff are seconded to PNICC to meet specific operational requirements, as required.[52] PNICC replaced the earlier National Reporting Centre (NRC),[53] created in 1972 to coordinate mutual aid requests from individual chief officers.

For the Games in 2012, there will be an additional coordination function operating from within a 'National Olympic Coordination Centre'. This multi-agency control room will be an additional resource and, at the time of writing, its location had not been agreed. Its function will be to manage the day-to-day policing activities of the Olympics and other events, supported by PNICC as and when necessary, and to respond to any major incident connected with the Olympics that might take place. All other day-to-day activity will be dealt with on a 'business as usual basis' by existing control rooms.

The police service within the UK is very experienced in cross-border mutual aid operations and nationally coordinated public order and serious crime investigations. The current system evolved gradually over time and was based on experience – and sometimes that experience was bitter. The experience of the various fire brigades in the UK was somewhat different. There have not been many incidents in recent times when a fire brigade has been overwhelmed by a particular incident; consequently, there was not an identified need to develop further capability. There have been some notable exceptions. The London Fire Brigade (LFB) developed the Inter-Agency Liaison Officer (ILO) scheme: a cadre of senior officers trained to work in a multi-agency environment and to understand the requirements of police incident commanders that would enable the LFB to provide a more informed response. The LFB also seconded a senior officer to work within the Anti-Terrorist Branch at New Scotland Yard. The individual fire brigades were, however, constrained by the Fire Service Act, 1947, which stated that their statutory duty was to make provision for fire fighting purposes.[54] There was no provision for rescue, floods, responding to road traffic accidents etc.

Chief Fire Officers used a supplementary power in a sub-clause to authorise the use of fire-brigade equipment and personnel to rescue cats up trees, as well as undertaking life-saving tasks, i.e. other than fire-fighting purposes for which it appears to the authority to be suitable.

Following the terrorist attacks on the United States on 11 September 2001, the UK government requested HM Chief Inspector of Fire Services to examine the Fire and Rescue Service's ability to respond to similar attacks. A 'New Dimension Group' was established, comprising Ministers and officials and coordinated by the Cabinet Office, to evaluate fire service capabilities and to make recommendations to ensure that it was sufficiently trained and equipped to deal with catastrophic, chemical, biological, radiological, nuclear and conventional terrorist incidents. By 2007, the New Dimension Programme was one of the key projects through which the government sought to build resilience, through a robust infrastructure response to deal rapidly, effectively and flexibly with the consequences of large-scale and catastrophic incidents – whether hazard (floods, pandemics, industrial disasters, etc.) or threat (terrorism).

The Fire and Rescue Service National Coordination Centre has been established to coordinate New Dimension equipment. The Department for Communities and Local Government (DCLG) is also developing an Emergency Room to facilitate the most effective deployment of Fire and Rescue Service/government equipment during any major emergency, in liaison with the National Coordination Centre.[55] Fire brigades became Fire and Rescue Services under the Fire and Rescue Services Act, 2004. This act replaced the outdated Fire Services Act, 1947. The new Act puts prevention at the heart of the Fire and Rescue Service by creating a new duty for all Fire and Rescue Authorities to promote fire safety, and provides powers to help create safer communities. The role of authorities in responding to road-traffic accidents and serious incidents such as major flooding and terrorist threats was also clarified.[56]

New Dimension equipment has been deployed to three major emergencies: Buncefield oil depot (December 2005), floods in the West of England and Yorkshire (summer 2007) and the Warwickshire warehouse collapse (November 2007). On each occasion, the results were encouraging.[57]

Nevertheless, in 2008, the National Audit Office review of the New Dimension Programme found some weaknesses in operational arrangements which hampered the maximisation of New Dimension effectiveness. The main areas of concern were Command and Control and major emergency planning. It pointed out that local commanders needed to know what to deploy, how to deploy it and to understand their powers. It recommended that DCLG should address any uncertainties with regard to National Coordination arrangements, by setting out clear frameworks, with respective roles and responsibilities for the Fire and Rescue Service National Coordination Centre and the Communities and Local Government (CLG) Emergency Room, and make these widely available.[58]

The closer cooperation between the police and fire services, particularly in the response to serious incidents such as terrorist attacks, explosions at oil storage facilities and flood disasters, is a useful indicator that many of the challenges to multi-agency coordination have been overcome, or are being actively pursued.

Planning

The coordination of an integrated approach to security is key to a safe and secure Games. The Olympic and Paralympic Policing Co-ordination Team (OPC) has been established to coordinate the delivery of police-led projects within the Olympic Safety and Security Programme. It is a joint MPS and ACPO team, but represents the police service nationally in a coordination role to ensure the police service is effectively geared to support the delivery of the London 2012 Games. Security planning for the Games is based on the wealth of experience held by all partner organisations within the Home Office's OSD, the OPC and other safety and security agencies.[59]

The March 2007 budget provided an additional £600 million for policing and wider security, tasks shared between a number of bodies including the Home

Office, Metropolitan Police, regional police forces and the emergency services. Although the Home Secretary is in overall charge of security for the Games,[60] at the operational level Metropolitan Police Assistant Commissioner (AC) Chris Allison was appointed on 8 May 2009, by the Metropolitan Police Authority,[61] to be the Assistant Commissioner for Central Operations, a role that will also encompass the position of National Olympic Security Coordinator at Games time. He is supported by the OPC and works closely with the Home Office's Olympic and Paralympic Security Directorate (OSD), led by Robert Raine, who has been in post since March 2008. The OSD is made up of staff from government departments including the Home Office and MOD, agencies including UKBA, staff on secondment from police services and other areas of expertise. Its role is to manage and oversee the Olympic Safety and Security Programme, and the Olympic Security Strategy that this Programme will deliver.[62]

In July 2008, Jonathan Stephens, Permanent Secretary, Department for Culture, Media and Sport, in oral evidence to the Committee of Public Accounts, stated that the Security Budget was £600 million, that all the planning is being done within that £600 million budget and he did not foresee any reason to expect the costed plan would depart from that.[63]

At the NHS Confederation's ambulance network conference in Harrogate, in July 2008, Chief Superintendent Stephen MacDonald from the Metropolitan Police Olympic Security Directorate identified some of the planning challenges that he and his colleagues were facing. For example, he confirmed that the £600m allocated to safety and security had to be shared by the Department of Health, Home Office and other departments such as transport. He said the figure was not a sum in a bank but a spending limit, meaning it was difficult for each department to know if it was overspending. He added that there would be little, if any, back-up from the armed forces. He also said that safety fears were so great there had been discussions about delaying the start of the football season for three months and cancelling the Notting Hill carnival.[64]

There has been no further public discussion on these two latter issues and it is highly unlikely that the Olympic Games will interfere with either the start of the football season or the Notting Hill Carnival. In October 2009, Lorraine Homer, the OSD Communication Manager, confirmed that Chief Superintendent MacDonald's comments reflected some of the very early thinking during planning for 2012. As would be expected in planning for an event of this size, there was a systematic process of understanding exactly what was required and planning how best to meet it. This meant that ideas and thinking had developed and would inevitably continue to develop over time.

She added that for some considerable time the OSD position had been that business as usual would continue. The police service anticipates a range of calls on their time in 2012 in addition to the Games, both in London and across the UK. The OSD strategy recognises this and the demand profiles for these events are being incorporated into the overall resource planning.

She concluded that the Ministry of Defence have been fully involved in security planning and work is going on to scope and understand what support the armed forces may provide, as they are prepared to do in other areas of occasional need.[65]

At the strategic level in London, the Metropolitan Police Authority (MPA) has been working to ensure that the governance and finances for the policing and security arrangements for the London 2012 Olympic and Paralympic Games are coherent and fit for purpose. They established the National Olympic Security Oversight Group, which brings together chairs of all the police authorities, fire authorities and ambulance trusts with 2012 venues in their areas. They have been an active member of the MPS management board's Olympics Oversight Group, the Home Office's Olympics Safety and Security Strategy Group and the ACPO Olympic Preparation Committee.[66]

Two new bodies have been set up. The Olympic Delivery Authority (ODA) will provide the facilities, and the London Organising Committee of the Olympic Games and Paralympic Games (LOCOG) will stage the Games. They are overseen by the Olympic Board, and a new team within the Department for Culture, Media and Sport (the Government Olympic Executive) will coordinate the contributions of other parts of government to the Games. The Olympic Board will play a leading role in progress monitoring and risk management, supported by a Steering Group of senior officials and the Olympic Programme Support Unit which provides independent advice to the Board. No one individual has overall responsibility for delivering the Games, however, and the large number of bodies involved presents significant risks, for example to timely decision-taking.[67]

Training and exercising

In the lead up to the Games in 2012, there will be a number of coordinated events designed to train personnel and test equipment, and a series of exercises to assess the effectiveness of contingency plans in an all-threats, all-hazards environment. These will be at local, regional and national levels.

The live events will be designed to assess the suitability of the venues for large crowds. These may include, in addition to medium- and large-scale sporting events, concerts and other such occasions that attract large audiences. The series of live events will test the facilities and event management in a benign environment, and the exercises will be designed to assist in the preparation of emergency service responders, and others, to deal with a variety of challenging disruptive events.[68]

The OSD intend to establish a small exercise coordination team. Following the 'commissioning' model, this team will be complemented by a more substantial exercise planning and delivery team that will sit in OPC. Overall, the OSD approach to Testing and Exercising focuses on ensuring that the 2012 baseline security measures and security operation is practised, prepared and ready for the Games. It will need to provide reassurance that the risks identified in the Olympic Safety and Security Strategic Risk Assessment (OSSSRA) have been adequately mitigated. In this context, testing will ensure that systems and processes work as intended whilst exercising focuses on exercising delivery structures and the responses to both routine events and reasonable worst-case scenarios. The Olympic difference necessitates a specific focus on exercising command, control and coordination; interoperability; information flows; and integration with 'business as usual' activity. The OSD are developing a Testing and Exercising strategy to provide a framework to ensure that these areas are addressed. Where the existing infrastructure does not readily accommodate a clear Olympic requirement, the OSD will commission and oversee the delivery of bespoke exercises.[69]

The Civil Contingencies Act 2004 (CCA) required all English regions and Wales to have a Generic Regional Response Plan (GRRP) enabling the activation of regional crisis management when needed. In London, the responsibility for strategic multi-agency planning lies with the London Regional Resilience Forum (LRRF). The LRRF is supported in its work by a number of additional groups including the London Resilience Programme Board, Sector Panels, Task & Finish Groups, Local Resilience Forums and the London Resilience Team. The LRRF developed the London Strategic Emergency Plan in response to the CCA requirement. The plan is tested through a multi-agency Regional Gold Exercise, staged approximately every 18 months, which is aimed at the strategic decision-making of the Partnership and the operation of the regional Gold Co-Ordination Group. The Partnership's Training and Exercising Strategy for 2010–2012 is being developed in conjunction with the Olympic Security Directorate and Civil Contingencies Secretariat during 2009 to better coordinate the London programme in the national and Olympics contexts.[70]

At the national level, the government and the emergency services regularly train and practise responses to a range of incidents, including malicious incidents, such as terrorism; natural disasters, such as floods and pandemics; and non-malicious human-made disasters, such as aircraft crashes, fuel-tanker fires, etc.

The Home Office leads and manages the UK's counter-terrorism exercise programme in cooperation with the Association of Chief Police Officers. This programme tests systems, trains frontline responders and highlights vulnerabilities. The national programme ensures that exercises reflect the changing nature of the terrorist threat worldwide. The programme has been running for over 30 years and forms an important part of counter-terrorism work. It is centred on three annual, large-scale live exercises with police forces around the UK, and other government departments and agencies to test counter-terrorist contingency plans. To support the large-scale exercises, a number of tabletop exercises are held in a variety of police regions.[71]

Public relations

During the period of the Games in 2012, the relationship between the public and the various authorities involved in managing the event, from government downwards, is anticipated to be one of mutual support. In the event of a terrorist or other challenging incident, it is essential to maintain this relationship through the provision of timely and accurate information. The research conducted by Donahue and Tuohy identified serious gaps in this relationship during the response to, and management of, major incidents. They found that the general public wants instructions about what to do, but that people may not receive or understand the directions government agencies give them. In part, this is because governments rely heavily on mainstream media. The research found that many people did not pay attention to mainstream media, and therefore did not get the information governments wanted them to have. Sometimes, those people who did get the information did not understand the message correctly, especially when the government gave little attention to pre-incident public education. This problem was exacerbated in the heat of an incident – when agencies failed to use a common message, did not control the message carefully, the pressure to get information out quickly undermined accuracy, and rumours propagated unchecked. Even when directions were clear, received and understood, some people did not have the wherewithal to follow them. The research found that some people just did not have the will to do as they were told and that the public was generally complacent about preparedness.[72]

The lessons for those responsible for managing a crisis response to any incident during the 2012 Games is to have a clearly defined, robust and clearly understandable media and public information strategy in place, well before the event.

Interoperability

The April 2009 HMIC report[73] into the implementation of the recommendations that followed the death of Mr de Menezes in July 2005 identified two core themes that had been repeatedly raised by the IPCC, debated at the Health and Safety trial, and investigated by the Coroner:

- MPS Counter-Terrorism Command and Control arrangements.
- MPS Counter-Terrorism Interoperability.

They were central to the events that led to the tragic shooting, and considered by HMIC to be ongoing matters of public confidence.

These themes will be a priority, not only for those planning the Games but also for those who will be engaged in the multi-agency safety and security operations in 2012. A comparative analysis of the HMIC model and the preparations to date is a worthwhile exercise.

HMIC Interoperability Model:

- Shared ethos (of what matters most).
    - *The shared ethos is to create a Games for all, where everyone is invited to take part, join in and enjoy the most exciting event in the world.*
- Common doctrine (a common set of operating principles or guidance).
    - *The guidance for a safe and secure Games is set out in the London 2012 Safety and Security Strategy.*

- A unified command model (clarity of who is in charge of what).
  - *The Safety and Security Strategy clearly identifies that the lead Government Minister is the Home Secretary, supported by a Metropolitan Police Assistant Commissioner, who is currently in charge of the operational planning for the Games and at the time of the Olympics will be the Security Coordinator. Each Chief Officer of Police will be responsible for Safety and Security for events taking place within their geographical boundaries.*
- Compatible and reliable communications systems.
  - *Airwave has been selected to be the emergency services radio system. Work continues to upgrade the system to meet anticipated demand.*
- Shared language (that ensures common understanding in pressurised operating environments).
  - *The emergency services have standard operating procedures for dealing with major incidents. However, there are still some areas for improvement. There is a national project to provide a common glossary for use by emergency service personnel.*
- Common equipment.
  - *Given the different missions of the emergency services, defining a common standard for equipment may not be achievable. Also at times of mutual aid, personnel from other services may not be familiar with bespoke equipment used by the organisation they are supporting.*
- Common standards of professional practice.
  - *National training programmes are intended to ensure consistency of professional practice. One area that continues to be a challenge is the collection, retention and dissemination of 'Organisational Learning'. The danger is that, if lessons are not learned, mistakes are likely to be repeated.*

This should be supported through a process of continuous capability building by:

- Shared training and exercising.
  - *The 2012 Games will have a regime of training and exercising. Given the numbers of emergency personnel and others who will be involved in the Games, it will be a challenge to ensure that those that need the training are trained and those who need to have the opportunity to exercise, attend a series of exercises.*
- Shared operational experience through working together.
  - *Emergency service personnel work together very effectively on a regular basis at Major Incidents across the UK.*
- Shared learning and debriefing.
  - *As mentioned above, the concept of organisational learning continues to be a challenge to the emergency services in general and the police service in particular.*[74]

- *Single agency debriefs are common, multi-agency debriefs are less common and where an incident/operation has 'gone wrong', the opportunity to debrief is postponed until after the variety of enquiries and legal processes (for example, the operation that led to the death of Mr de Menezes was not formally debriefed until four years after the event).*[75]

There are some positive results coming from the analysis of the HMIC interoperability model and the work in progress by those implementing the 2012 Games Safety and Security Strategy. The work is not complete, but it is encouraging to note the close correlation between the HMIC model and the manner in which the Olympic and Paralympic Games Safety and Security Strategy is being implemented.

Conclusion

Preparations and planning for the safety and security of the 2012 Olympic and Paralympic Games is well under way. Responsibility for safety and security rests with the Home Office, which has identified that safety and security is everyone's responsibility, not just the police and security and intelligence agencies. Lord West has listed those departments, agencies and organisations that have a role to play in ensuring a safe and secure Games (see Appendix). A Safety and Security Strategy has been published and operational planning is being undertaken to ensure that those who need to be trained and exercised will receive the support they require. As the various venues are being built, capabilities are being enhanced. The intended outcome, in the event of a Major Incident during the period of the Games – including malicious incidents, such as a terrorist attack; or a natural disaster; or a non-malicious, human-made disaster – is that the necessary processes, procedures and measures needed to respond to, mitigate against and recover from the incident should be in place. There is no doubt that the intrinsic challenges of multi-agency coordination will be multiplied many times by the magnitude of the Olympics.

The success of the wise investment in planning, training, exercising, resourcing, managing and coordinating activity, in the lead up to the Games, will be measured by the effectiveness and efficiency of the response to any disruptive event, malicious or otherwise, of the emergency service personnel who will be on duty during the period of the Olympic and Paralympic Games in 2012.

Appendix

The list of departments, agencies, public bodies and other organisations that participate in or contribute to the Olympic safety and security programme. (Identified by Lord West in January 2009, in answer to a question from Baroness Neville-Jones.)

1 Association of Chief Police Officers (ACPO)
2 Association of Police Authorities (APA)
3 British Red Cross
4 British Telecom
5 British Transport Police (BTP)
6 British Transport Police Authority
7 Cabinet Office
8 Centre for the Protection of National Infrastructure (CPNI)
9 City of London Police
10 City of London Police Committee
11 Department for Communities and Local Government (DCLG)
12 Department for Culture, Media and Sport (DCMS), including
   a the Government Olympic Executive (GOE)
13 Department for Health (DoH), including
   a the wider National Health Service (NHS)
   b the Health Protection Agency (HPA)
14 Department for Transport (DfT), including
   a Transport Security and Contingencies Team (Transec)
15 Equalities and Human Rights Commission
16 Foreign and Commonwealth Office (FCO)
17 Greater London Authority (GLA)
18 Government Office for London (GO – London)
19 Government Office for Science (GOS)
20 Health and Safety Executive (HSE)
21 Her Majesty's Inspectorate of Constabulary (HMIC)
22 HM Courts Service
23 HM Treasury, including
   a Office of Government Commerce (OGC)
24 The Home Office, including
   a Criminal Records Bureau (CRB)
   b Home Office Scientific Development Branch (HOSDB)
   c Identity and Passport Service (IPS)
25 National Policing Improvement Agency (NPIA)
26 Security Industry Authority (SIA)
27 Serious Organised Crime Agency (SOCA)
28 International Olympic Committee (IOC)
29 London Ambulance Service and ambulance services at other venues
30 London Boroughs
31 London 2012 Organising Committee (LOCOG) and UK and international sports organisations
32 London Criminal Justice Board
33 London fire brigade and fire and rescue services at other venues
34 Maritime and Coastguard Agency
35 Metropolitan Police Service (MPS)
36 Metropolitan Police Authority (MPA)

37 Ministry of Defence (MoD)
38 Ministry of Justice (MoJ) and related criminal justice agencies
39 National Grid
40 Office of the Chief Scientific Officer
41 Olympic Delivery Authority (ODA)
42 Other Home Office police forces, including those responsible for policing at specific venues, such as
    a Dorset
    b Essex
    c Greater Manchester
    d Hertfordshire
    e Northumbria
    f South Wales
    g Thames Valley
    h West Midlands, or related facilities
    i Kent
    j Surrey
    k Sussex
43 Port of London Authority (PLA)
44 Private companies and trade associations, including the British Security Industry Association (BSIA)
45 The Security Service
46 Strathclyde Police
47 Transport for London (TfL)
48 Transport operators (railways, ports and airports)
49 United Kingdom Border Agency (UKBA).

Notes

1. http://security.homeoffice.gov.uk.
2. www.olympic.org/uk/organisation/missions/truce/truce_uk.asp.
3. Home Office (2009) *Olympic and Paralympic Safety and Security Strategy*, July.
4. www.london2012.com/about/our-brand/index.php.
5. http://inspectorates.homeoffice.gov.uk/hmic/inspections/special_humberside_police_report/.
6. www.mpa.gov.uk/downloads/scrutinites/stockwell/hmic-mpsprogress.pdf.
7. www.ipcc.gov.uk/stockwell_one.pdf.
8. www.mpa.gov.uk/downloads/committees/mpa/080724-06-appendix01.pdf.
9. www.stockwellinquest.org.uk/hearing_transcripts/DeMenezesRule43Reportopen.pdf.
10. Home Office (2009) *London 2012. A Safe and Secure Games for All*.
11. Home Office (2009) *Olympic and Paralympic Safety and Security Strategy*, July.
12. 'Severe – an attack is highly likely', www.cabinetoffice.gov.uk/security_and_intelligence/community/threat_levels.aspx.
13. Home Office (2009) *Olympic and Paralympic Safety and Security Strategy*, July, p. 11.
14. The Joint Terrorism Analysis Centre, or JTAC, was created as the UK's centre for the analysis and assessment of international terrorism. It was established in June 2003 and is based in the Security Service's headquarters at Thames House in London. www.mi5.gov.uk/output/joint-terrorism-analysis-centre.html.

15 'The current threat level is assessed as Substantial (as of 20th July 2009).' 'Substantial – an attack is a strong possibility', www.cabinetoffice.gov.uk/security_and_intelligence/community/threat_levels.aspx.
16 Homer, L. (2009) Home Office Olympic Security Directorate, email: 7 October.
17 Home Office (2009) *Olympic and Paralympic Safety and Security Strategy*, July, p. 6.
18 Home Office (2009) *Olympic and Paralympic Safety and Security Strategy*, July, p. 6.
19 Home Office (2009) *Olympic and Paralympic Safety and Security Strategy*, July, p. 8.
20 Home Office (2009) *Olympic and Paralympic Safety and Security Strategy*, July, p. 9.
21 Jennings, Will (2005) 'London 2012 and the risk management of everything Olympic', *Risk and Regulation* 10 (winter), p. 9, www.lse.ac.uk.
22 Jennings, Will (2005) 'London 2012 and the risk management of everything Olympic', *Risk and Regulation* 10 (winter), p. 9, www.lse.ac.uk.
23 Home Office (2009) *Olympic and Paralympic Safety and Security Strategy*, July, p. 9.
24 Alison, Laurence and Crego, Jonathan (2008) *Policing Critical Incidents*, Devon: Willan Publishing, p. xxvii.
25 Alison, Laurence and Crego, Jonathan (2008) *Policing Critical Incidents*, Devon: Willan Publishing, p. xviii.
26 Lord West (2009) House of Lords, 22 January, http://services.parliament.uk/hansard.
27 The Senior Responsible Owner (SRO) is the individual responsible for ensuring that a project or programme of change meets its objectives and delivers the projected benefits. They should be the owner of the overall business change that is being supported by the project. The SRO/PO should ensure that the change maintains its business focus, has clear authority and that the context, including risks, is actively managed. This individual must be senior and must take personal responsibility for the successful delivery of the project. They should be recognised as the owner throughout the organisation. www.ogc.gov.uk/User_roles_in_the_toolkit_senior_responsible_owner.asp.
28 Homer, L. (2009) Home Office Olympic Security Directorate, email: 7 October.
29 Donahue, Amy and Tuohy, Robert (2006) 'Lessons we don't learn: a study of the lessons of disasters, why we repeat them, and how we can learn them', *Homeland Security Affairs* II, 2, July.
30 Donahue, Amy and Tuohy, Robert (2006) 'Lessons we don't learn: a study of the lessons of disasters, why we repeat them, and how we can learn them', *Homeland Security Affairs* II, 2, July, p. 22.
31 Donahue, Amy and Tuohy, Robert (2006) 'Lessons we don't learn: a study of the lessons of disasters, why we repeat them, and how we can learn them', *Homeland Security Affairs* II, 2, July, p. 5.
32 ACPO NPIA (2009) *Guidance on Multi-Agency Interoperability*.
33 Alison, Laurence and Crego, Jonathan (2008) *Policing Critical Incidents*, Devon: Willan Publishing, p. 236.
34 7th July Review Committee (2006) Greater London Authority, www.london.gov.uk.
35 London Assembly (2007) 7 July Review Committee, follow-up report, August 2007, www.london.gov.uk.
36 www.npia.police.uk.
37 www.npia.police.uk.
38 Terrestrial Trunked Radio (TETRA) comprises a suite of open digital trunked radio standards used by Private Mobile Radio users such as those involved in public safety, transportation, utilities, government, commercial and industrial, oil and gas, and military etc. TETRA is an interoperability standard that allows equipment from multiple vendors to interoperate with each other. www.tetramou.com/.
39 www.publictechnology.net/modules.
40 London Assembly (2007) 7 July Review Committee, follow-up report, August 2007, p. 18, www.london.gov.uk.
41 Donahue, Amy and Tuohy, Robert (2006) 'Lessons we don't learn: a study of the

lessons of disasters, why we repeat them, and how we can learn them', *Homeland Security Affairs* II, 2, July, p. 6.
42 Home Office (2009) *Olympic and Paralympic Safety and Security Strategy*, July, p. 17.
43 London Emergency Service Liaison Panel (2007) *Major Incident Procedures Manual* (7th edition).
44 Survey/Safety; Assess; Disseminate/Declare; Casualties; Hazards; Access; Location; Emergency Services (range and commitment); Type of incident.
45 The terms 'Gold', 'Silver' and 'Bronze' have not been adopted by the emergency services in Scotland. See *Preparing Scotland, Scottish Guidance on Preparing for Emergencies* (2007), p. 30.
46 London Emergency Service Liaison Panel (2007) *Major Incident Procedures Manual* (7th edition).
47 Arbuthnot, Kevin (2008) 'A command gap? A practitioner's analysis of the value of comparisons between the UK's military and emergency services' command and control models in the context of UK resilience operations', *Journal of Contingencies and Crisis Management* 16, 4, p. 192.
48 Supply chain: 'the network of organizations that are involved, through upstream and downstream linkages, in the different processes and activities that produce value in the form of products and services in the hands of the ultimate consumer' (Christopher, M. (1992) *Logistics and Supply Chain Management*, Pitman Publishing, London). See Christopher, Martin and Peck, Helen (2004) 'Building the resilient supply chain', *International Journal of Logistics Management* 15, 2, pp. 1–14.
49 Home Office (2009) *Olympic and Paralympic Safety and Security Strategy*, July, p. 13.
50 Interview with DAC Richard Bryan, New Scotland Yard, 29 July 2009.
51 While there is no UK-wide policy specifically relating directly to mutual aid, many areas such as police, fire, NHS and local authorities have inter- (and intra-) agency mutual aid protocols in place. Most of these are formal, but many are informal. Formal protocols detail how each partner will undertake or allocate responsibilities to deliver tasks. Protocols may cover matters of broad agreement or details for working together, including how to hand over tasks or obtain additional resources. Protocols may or may not be legally binding depending upon the nature of the agreement between the parties. www.cabinetoffice.gov.uk/ukresilience/response/recovery_guidance/generic_issues/mutual_aid.aspx See also *Mutual Aid, Guidance for Local Authorities* (2008) www.cabinetoffice.gov.uk/media/132859/mutual_aid.pdf.
52 Hanson, David (2009) Minister of State (Crime and Policing), Home Office, 21 July, www.publications.parliament.uk/pa/cm200809/cmhansrd/cm090721/text/90721w0093.htm#09072311010387.
53 Its main purpose was to help in the national coordination of aid between Chief Police Officers in England and Wales, under section 14 of the Police Act 1964, so that the best use was made of manpower and to provide the Home Secretary with relevant information. http://hansard.millbanksystems.com/written_answers/1984/apr/05/police-national-reporting-centre.
54 Fire Services Act, 1947, S1(1), www.uklegislation.hmso.gov.uk/RevisedStatutes/Acts/ukpga/1947.
55 'New Dimension – enhancing the Fire and Rescue Services' capacity to respond to terrorist and other large-scale incidents', 31 October 2008, National Audit Office, www.nao.org.uk.
56 http://uk.ihs.com/news/newsletters/ohsis/ohsis-nov04-fire-rescue-act-2004-odpm-factsheet.htm.
57 'New Dimension – enhancing the Fire and Rescue Services' capacity to respond to terrorist and other large-scale incidents', 31 October 2008, National Audit Office, www.nao.org.uk, p. 21.

58 'New Dimension – enhancing the Fire and Rescue Services' capacity to respond to terrorist and other large-scale incidents', 31 October 2008, National Audit Office, www.nao.org.uk, p. 9.
59 Homer, L. (2009) Home Office Olympic Security Directorate, email: 7 October.
60 www.publications.parliament.uk/pa/cm200708/cmselect/cmpubacc/85/8505.htm#n11.
61 www.whitehallpages.net.
62 Homer, L. (2009) Home Office Olympic Security Directorate, email: 7 October.
63 Stephens, Jonathan (2008) Permanent Secretary, Department for Culture, Media and Sport, *Preparations for the London 2012 Olympic and Paralympic Games: Progress Report June 2008*, www.publications.parliament.uk/pa/cm200708/cmselect/cmpubacc/890/890.pdf.
64 Santry, Charlotte (2008) *Health Service Journal*, 3 July.
65 Homer, L. (2009) Home Office Olympic Security Directorate, email: 14 October.
66 Metropolitan Police Annual Report (2007/2008) www.met.police.uk, p. 43.
67 House of Commons Committee of Public Accounts (2007) *Preparations for the London 2012 Olympic and Paralympic Games – Risk Assessment and Management*, 27 June, www.publications.parliament.uk/pa/cm200607/cmselect/cmpubacc/377/377.pdf.
68 Interview with Chief Superintendent Jim Busby, Metropolitan Police, 9 September 2009.
69 Homer, L. (2009) Home Office Olympic Security Directorate, email: 7 October 2009, citing Commander Ian Quinton OSD.
70 www.londonprepared.gov.uk/downloads/emergplanv5.pdf.
71 http://security.homeoffice.gov.uk/responding-terrorist-incident/national-response/exercise-programme/.
72 Donahue, Amy and Tuohy, Robert (2006) 'Lessons we don't learn: a study of the lessons of disasters, why we repeat them, and how we can learn them', *Homeland Security Affairs* II, 2, July, p. 6.
73 http://inspectorates.homeoffice.gov.uk/hmic/inspections/special_humberside_police_report/.
74 Progress in drawing up the firearms and surveillance glossaries is advanced but will require national negotiation and agreement prior to implementation. www.mpa.gov.uk/downloads/scrutinites/stockwell/hmic-mpsprogress.pdf.
75 With the conclusion of the inquest into the death of Jean Charles de Menezes in December 2008, and the publication of the Coroner's Rule 43 report in March 2009, the MPS has begun a formal and structured debrief involving key groups and individuals (and identified IPCC staff) to learn as much as possible from all the events of July 2005. A date for completion has yet to be determined. The Specialist Firearms Command (CO19) is not taking part because the MPS has decided that many of the issues for them have already been identified during the Health and Safety trial and the Inquest. www.mpa.gov.uk/downloads/scrutinites/stockwell/hmic-mpsprogress.pdf.

11 The European Union and the promotion of major event security within the EU area

Frank Gregory

Introduction

The EU has a long history of regarding aspects of public-order maintenance as a matter of common concern going back to the post-1975 TREVI (pre third pillar JHA, TEU) system on cooperation, which included a focus on football hooliganism intelligence and policing. Football hooliganism was then a significant major event security problem, as indeed it continues to be, albeit more sporadically. Domestic or national policing responsibilities remain unchanged except in respect of obligations to cooperate, share 'information' and expertise. However, the Amsterdam Treaty introduced the wider aim of making the EU an 'area of freedom, security and justice' – this implies that wherever an individual is in the EU, they should enjoy a common standard of security at a 'major event' (on JHA general issues see, for example, Crawford, 2002; Monar *et al.*, 2003; Walker, 2004; Guild and Geyer, 2008). This point was specifically referred to by the EU Council, in May 2004, in the context of the Athens Olympics: 'The EU's objective is to provide citizens with a high level of safety within an area of freedom, security and justice by developing common action among the Member States in the field of police co-operation' (Council of the EU, ENFOPOL 14, 2004). The EU had, of course, already experienced the major Olympics security incident of the attack on the Israeli team at the 1972 Munich Olympics. Post-9/11, major events, whether political, cultural or sporting, were highlighted by the EU as possible iconic targets for terrorists. It is important to note that, whilst EU event security manuals may be said to reflect a basic common understanding of 'security' related to the organisation of major events, 'security' still remains a contested concept among EU Member States (ESRIF, 2009: 213–219; and see also Edwards and Hughes, 2005).

During the development of what is now the 2009 Lisbon Treaty (which represents the latest stage of the evolution of EU integration), the Amsterdam Treaty's aspirations in the JHA area were also further developed by the Council of Ministers via the 'Stockholm Programme – An Open and Secure Europe Serving and Protecting the citizens'. This is seen as both complementary to and an integral part of the 2003 European Security Strategy. The Spanish Presidency (January–June 2010) has sought to further the Stockholm Programme by developing 'a multidimensional Internal Security Strategy' (Gruszczak, 2010). Among the principles being set out for this proposed ISS, the following are applicable to major event security (MES):

- the need for a horizontal and cross-cutting approach in order to be able to deal with complex crises or natural or man-made disasters (this latter category includes terrorism);
- reflecting a proactive and intelligence-led approach.

This EU-level objective creates both obligations and expectations in respect of the delivery of policing as a public good in EU states but, additionally, because many major events are both private sector ventures and held on private spaces, the objective raises questions about the event security roles of the private sector including the private security industry. This objective is reinforced by some MES criteria being specified by international sporting bodies. For example, in the area of MES, Body-Gendrot noted (2003: 40) that the traditional French police monopoly in the management and delivery of MES had to be modified because of the increasing burden of safety provision in urban spaces which requires public–private cooperation, the influence of international bodies (e.g., the EU and FIFA) on security policy for 'mega events' and the greater involvement of local government in MES. However, Ocqueteau (2006: 73) noted that the French government only allowed private sector involvement in MES 'after a very meticulous definition of powers provided to private agents, a necessary condition in the face of the state agents' reserved opinions about extending the role of private security'.

Examples of EU MES policy development can be found in EU sources such as the Joint Action (97/339/JHA) regarding cooperation on public order and security (May 1997) and the JHA Council Conclusions (10916/01JAI82) on security at meetings of the European Council and other public events (July 2001). Such policies, prior to the 2009 Lisbon Treaty, were discussed in JHA Council expert groups, including police cooperation networks, and these discussions fed into the Council General Secretariat Security Office which advises the Council of Ministers on security. Under the Lisbon Treaty, the EU Council supervisory body for JHA/ISS will be the Internal Security Committee (COSI). The European Commission plays quite a wide-ranging role in relation to MES because of its responsibilities in the area of health and safety and its economic area responsibilities for the promotion of both the development of the European security equipment and systems industry and the private security industry. The Security Office of the European Commission can also be involved 'where necessary' (Council of the EU, ENFOPOL 123, 2002). EU MES policy documents set out a number of important principles governing the delivery of event security.

In the *Security Handbook for the Use of Police Authorities and Services at International Events Such as Meetings of the European Council* (EU, ENFOPOL 123, 2002), two public order principles are set out:

1 'The enforcement of law and order should be guided by the principles of proportionality and moderation preferring the less intrusive approach. Where possible, a deescalating **police** approach should be chosen.'
2 'Dialogue and cooperation with demonstrators and activists should be actively pursued by the **police authorities**.'

The implementation requirement is for each Member State to have a 'national contact point' which is supposed to: collect, exchange and disseminate information and risk analyses. Under the December 2006 EU Presidency Proposal for a single *Security Handbook for the Use of Police Authorities and Services at International Events* (which leaves the 2006 EU football policing handbook as a 'stand alone' document), the Basic Principles include the note:

> Although the host Member State has primary responsibility for providing for the security at the event, given its international character, all other Member States and EU competent bodies have a responsibility to assist and support the provision of such security.
> (Presidency of EU Council to Police Cooperation Working Party, ENFOPOL 190, 2006)

The *Handbook for Member States Co-Operation Against Terrorist Acts at the Olympic Games and Other Comparable Sporting Events* is considered to be 'an evolving instrument' for MES which needs to be updated via future experiences and best practice. It specifically needs inputs on: assessments of terrorist threats, suspect persons and threat levels (Council of the EU, ENFOPOL 14, 2004). Similarly, in the EU Presidency note *Proposals Relating to the Enhancement of Measures to Counter Football Related Violence* (EU, ENFOPOL 23, 2004), there are also to be found suggestions as to improvements that could be made to the *Football Handbook* such as:

1 improving the operational use of categories of estimated risk;
2 better information on Member States' travel restrictions rules;
3 using the annual ad hoc report on football vandalism compiled by Belgium, the UK, the Netherlands and Germany;
4 mutual assessments of police effectiveness at major football events; and
5 developing a website because 'A single resource and reference centre for police working in this area could be very helpful'.

It is important to establish, at an early stage in this analysis, the constraints upon the EU in both policy-making and policy implementation (on these general points, see, Bache and George, 2006). First, under both treaty provisions and the principle of 'subsidiarity', the EU has no direct powers over either the provision or the delivery of internal security within Member States. Second, even where there is provision for the EU to provide assistance in, for example, MES, by comparison with the Member States the EU has very limited physical and budgetary resources. Third, because of this lack of 'own resources', the EU is dependent for much of the implementation of agreed policies on the implementing actions of Member States. Unfortunately, Member States have a variable record in this regard. Fourth, in the area of MES, the guidance manuals are not specific EU law measures with primacy over domestic law but, rather, weaker measures under general EU treaty obligations. Thus, Member States have a choice with respect to EU policy on MES: they can implement the manuals' guidance in full in their national MES policies or they can simply 'take note' of the advice. However, it has been cogently argued (Jennings and Lodge, 2009) that such sources may form part of an institutional isomorphism and could be part 'of a dominant or hegemonic discourse regarding the appropriate tool of security risk management'. Member States can also make full use of the relevant services of the Commission and Europol, and/or they can seek assistance through intergovernmental channels and from the various police, security and intelligence and other emergency response networks, and from the private sector. These options are well illustrated by the Greek security organisational response to the 2004 Athens Olympics, which is discussed later (pp. 216–217).

Since 2004, the European Commission has been funding, under FP6, a specific major EU project on security at major events known as EU–SEC (Europa, 2006). The first phase involved the police and/or interior ministries of ten EU MS, Europol and the United Nations Interregional Crime and Justice Research Centre (UNICRI), and focused on establishing 'best practice' in MES and, in particular, the opportunities for sharing and cooperating in the area of research on MES. UNICRI is involved because it maintains the UN's International Permanent Observatory on Security Measures during Major Events. The second phase of EU–SEC (EU–SEC II), which is currently ongoing, has a focus on public–private partnerships (PPPs) in MES. The 'value-added' of the EU–SEC project is, as yet, not easily discernable. There have been two particular problems with the EU–SEC project, in both its phases: first, the participating EU Member States have a variable record in sending representatives with appropriate major event security experience and, second, the participating states also have a variable record in responding to agreed requests for national major event-related inputs. All of which leaves the EU Commission's aspiration for developing EU MES 'best practice' reliant more on a negative assent process rather than always being based on well-grounded, evidence-based positive assent by the participating states. To date, EU–SEC has produced a manual on *Coordinating National Research Programmes on Security During Major Events in Europe*, a restricted portal for information sharing, a common database on existing MES research and a report on ethical issues at MES (European Commission, 2008a).

This chapter will use the above general grounding in the background to the EU's approach to major event security to develop an analysis of EU MES policy via examinations of the football major events security system, general information sharing on security concerns related to major events, the specifics of EU Olympics security policy and the issues surrounding public–private partnerships in MES.

MES – international football games: the football policing network

The most highly developed area of EU MES cooperation relates to football matches and series with an international dimension. This long-established, high-profile network derives its cohesion and commitment to sharing information and research from the following factors: the popularity, national prestige and private sector investments (sponsorships, etc.) attached to football events; the need to ensure that a country is seen as having appropriate stadium standards and good behaviour by its football fans in order to not incur FIFA or UEFA bans; the common EU aim to prevent football hooliganism; the clear lead from the EU Council by its December 2001 Decision that all Member States must designate a single national point of contact (NFIPs) for football policing issues related to European or international games; and that its expertise has a 'transferability' value for other major public events through the dissemination of best practice. For example, the Italian police football intelligence officer was also involved in the security arrangements for the 2006 Winter Olympics in Turin.

This sector does not report any significant problems in cooperation other than the variable capabilities of the resources available for national contact points. For example, the UK 'national point of contact', the UK Football Policing Unit, is supported by 92 police football intelligence officers ('spotters') located in the constituent forces of the UK police system. There is one such officer for each major league football club. Some other EU states do not have such a large support network for their national point of contact. However, not surprisingly, there are particular and variable national constraints in the sharing of personal data under both data protection and human rights legislation. Additionally, information or research that might be derived in whole or in part from national intelligence sources, such as a UK JTAC (Joint Terrorism Analysis Centre) Assessment would only be shared in a suitably indirect manner and on a strictly 'need to know' basis.

This network commissions academic research into areas such as football hooliganism, crowd behaviour control strategies and risk management. Moreover, countries hosting very high-profile football events make significant efforts to promote information sharing. For example, in preparing for the 2006 World Cup, the German Federal Ministry of the Interior organised international conferences in Berlin in 2002 and 2003 to share experiences on major sporting events. Additionally, German fire department officials provided a detailed discussion of World Cup related CBRN measures, in conformity with the Nationales Sicherheitskonzept, to the professional journal *NBC International*, (Winfield, 2006: 72–75).

The network is also developing the peer review process, as proposed in ENFOPOL 23 in 2004, and, in due course, the result of the peer reviews will be disseminated. The website, mentioned in ENFOPOL 23, has been constructed and is available as the European National Football Information Point website, under password control for NFIPs and Europol, on the UK Centrex (police central training facility) website. This website contains not only police originated information for sharing, but academic research papers as well.

This network benefits from the fact that the high political visibility of international football events linked to their relative frequency of occurrence has ensured that football policing cooperation issues are a standing agenda item for each EU Presidency based on a *Report from the Police Cooperation Working Party* (experts on major sporting events). The network also makes full use of the impetus that can be provided by an active 'core group' of Member States that routinely facilitates activities through the close cooperation of the Presidency and adjacent near-past and near-future Presidency states.

Information/research sharing networks relating to public order, terrorism and extremism

Unlike the more bounded working environment of the EU football hooliganism intelligence network, the non-football European major events public order intelligence network operates within a more diverse working environment. This can be summarised by the following points: the location and frequency of events such as EU Council meetings, state and VIP visits, G8 Summits and Olympic Games are more varied than with major football events; the likely public order problems are essentially political in character and, apart from terrorist threats, grounded in the democratic right of public protest as opposed to the mindless violence of football hooligans; EU Member States have a variable experience with the diverse 'protest groups' according, in part, to national circumstances; the 'protest groups' cover a broad spectrum of single-issue and multi-issue concerns, for example, EU farmers' groups protesting against an aspect of CAP reform, animal rights groups, environmental protection action groups, anti-capitalist groups and right- or left-wing political extremist groups. The only common theme, from a public order policing perspective, is that of 'extremism', which implies a willingness to use violence against persons or property. Consequently, outside the football area, the police information/research sharing networks are more diverse in character.

Unlike the football intelligence network, which has clear EU institutional linkages, the public order network is based upon the wider membership Police Working Group on Terrorism (PWGT), an inter-agency network with governmental recognition. PWGT utilises both its own secure communications network and links into a network of national liaison officers known as CTELOs (Counter-Terrorism and Extremism Liaison Officers). For example, the UK CTELO in France is attached to UCLAT in Paris, and the French CTELO to the UK is located within the Metropolitan Police Counter-Terrorism Command (SO15).

Under the TEU, as amended by the Treaty of Amsterdam, terrorism is defined as both a matter of 'common interest' (Article 29) and as a threat to the achievement of making the EU an 'area of freedom, security and justice' (Article 2). However, when the CTELO network is tracked back to 'national contact points' in the EU Member States, it is evident that the terrorism aspects are handled in a separate manner from other public order issues. For example, in the UK the National Coordinator for Domestic Extremism (NCDE) oversees the work of the National Public Order Intelligence Unit (NPOIU), but the NPOIU only feeds into the CTELO network on non-football and other major event security issues on issues related to extremists and extremist groups. Police information and intelligence related to terrorism flows from the UK to the European partners through the non-EU PWGT via the International Section of the Metropolitan Police Counter-Terrorism Command (SO15). Well before 9/11, and even before the activation of Europol's counter-terrorism mandate in 1999, there were occasional attempts at the Europeanisation of counter-terrorism information flows. For example, in 1996 a JHA Joint Action had required the creation and maintenance of a 'Directory of Specialised Counter-Terrorist Competencies, Skills and Expertise in the Member States'. However, that JA has not really been fully and continuously implemented owing to the variability of responses from Member States.

Whilst Member States always have, and will continue to seek, appropriate assistance in response to counter-terrorism threats and/or incidents in relation to MES – as did Greece in relation to the 2004 Athens Olympics – they do not seek assistance from other EU states with regard to public order problems that may arise during an ME (except, of course, for intelligence on extremists or extremist groups and intelligence on football hooligans) because public order maintenance is purely a national responsibility. This is still the case, even where there are significant and controversial public order problems, as in the cases of the 2001 Genoa G8 Summit, the 2001 Gothenburg Summit, the 2007 Heiligendamm G8 Summit and the 2009 London G20 Summit. Following the 2001 ME public order problems, Germany did propose that the EU consider the creation of 'Special units to guarantee the safety of meetings of the European Council and other comparable events' (Council of the EU, ENFOPOL 96, 2001). Although this proposal was not acted upon, its details are set out below as they merit consideration within this analysis.

The justification for proposing common action in relation to public order problems was similar to that advanced over common action against football hooligans. The preamble to ENFOPOL 96 argued that EU citizens have a right to conduct peaceful protests in Member States in a safe and secure manner, but that this right as 'shown by the events in Göteborg and Genoa ... will be possible only if the Member States of the European Union implement joint and harmonized measures against traveling offenders committing violent acts'. The key point, in EU policy-making terms, is the reference to 'traveling offenders committing violent acts', i.e. a problem that affects two or more Member States and crosses intra-EU borders.

The proposal has two elements: first, the 'creation of common standards for training and equipment of existing special units in Member States ... entrusted with the task of ensuring public safety and order ...' (ibid.: 2).[1] This is a reference to units such as the French CRS and the German Bereitschaftspolizei. In parts, this proposal is not too controversial where it refers to a 'common tactical framework concept to ensure the safety of meetings of the European Council and other comparable events ...' (ibid.). The suggestion is that this framework could be derived from institutionalising public order policing experience exchanges, joint exercises and exchanges of information on national legal regimes. However, the proposal also refers to equipment standardisation, including in the area of weapons. This would be a controversial element as some EU Member States have tear gas and water cannons available, and their police routinely carry firearms. The second element of the proposal was that 'the preconditions must be established to enable one Member State to request the support of special units from other Member States ...' (ibid.: 3). Not only was this clearly a controversial measure and not accepted, but it implied that a Member State would be willing to concede that it could not carry out one of the basic duties of a sovereign government, that of maintaining internal order. However, although the German initiative was unsuccessful, Member States, through police and intelligence networks, do of course share both intelligence and experience with respect to the provision of security at political summits held in EU states. It should be noted, though, that a later move in 2004 by an EU Council Resolution to encourage Member States to share information on people travelling to protest was felt, by the House of Lords' EU Committee (HL 119, 2004), to raise civil liberties implications.

EU and Olympics security

The Olympics are held every four years and, of course, the venue choices rotate around the major geographical divisions of the world. Therefore it is not a major event category that forms part of the regular annual cycle of MEs held within the EU area – for example, there was the 2004 Athens Olympics, the 2006 Turin Winter Olympics and there will be the 2012 London Olympics. However, the scale and prestige attached to the Olympics and the fact that all the EU Member States participate make the Olympics, when held in an EU state, a natural focus for EU cooperation in MES. Nonetheless, nothing in the EU treaties removes from the Olympics host Member State the ultimate responsibility for internal security. As discussed earlier, the EU has produced generic guidance for MES which has applicability to the Olympics; and, whilst this encourages appropriate inter-EU cooperation, it still reinforces the fact that the host state and whatever national entity is deemed to be the Olympics organiser have the primary duty to provide security for the Games.

The post-9/11 heightened awareness of the threat of international terrorism to the EU area, and the successful Greek bid for the 2004 Olympics, led the EU Council to draw up a *Handbook for Member States' Co-Operation Against Terrorist Acts at the Olympic Games and Other Comparable Sporting Events* (EU, ENFOPOL 14, 2004). The EU Council noted that, between 2004 and 2007, there were the following major sporting events in the EU: the Athens Olympics, the Turin Winter Olympics, the America's Cup, the Mediterranean Games and the World Athletics Championship. Because of the Amsterdam Treaty commitment to providing '[EU] citizens with a high level of safety ... by developing common action ...' (ibid.), the EU Council's view was that one way of meeting that commitment was for there to be a relevant EU handbook (to be considered as an evolving and updateable instrument), available for the organising national authorities, developed through the usual EU security cooperation fora. In this instance, contributions from the Police Chiefs' Task Force were particularly noted. The handbook's key elements are:

- all EU MS and EU 'competent bodies' have a duty to assist and support the provision of MES by the organising state;
- that the organising state must draw up an 'updated threat assessment and risk analysis six months before the relevant event at the latest, updateable until the conclusion of the event using all available national and EU sources' (e.g., Europol);
- that the organizing MS will select 'suitable, necessary and appropriate security measures ...' based on the risk analysis;
- that standardised EU communications systems, especially using permanent national contact points in MS, should be used for information exchange on the prioritised categories of: terrorist suspects and suspicious activities, threat levels against VIPs, athletes, visitors/spectators and venues;
- minimum standards for national resources to be available to permanent national contact points;
- provision for the Olympic Games host MS to request other MS to appoint liaison officers.

More specifically, the Olympics host MS is expected to:

- clearly and publicly set out the roles of all the relevant authorities and services in respect of the Games;
- draw up a comprehensive communications plan between its Olympics ME management structure and all the 'key players', e.g. the IOC, team-contributing states, etc.;
- operational planning – the organising MS must draw up: strategic, operational and tactical plans;
- the organising Member State may 'request the deployment of police or intelligence officers for operational support from another Member State for the fight against specific terrorist scenarios ...' and may 'arrange for support from other Member States when possible through bilateral/multilateral agreements on temporary provision of equipment or other resources ...'

Some evidence of a host Member State implementing this guideline format can be discerned from the case of Greece and the 2004 Athens Olympics (see Migdalovitz, 2004). Greece provided a $1.2bn budget for Olympic security, designated the Ministry of Public Order as the lead body for Olympic security (its Minister chaired the inter-ministry Coordinating Council for Olympic Security) and deployed 40,000 police, and 10,000 soldiers (200 with CBRN incident-response training – helped by IAEA advice). The Greek Olympics planning phase started in 2000 and included the establishment of a seven-nation Olympic Advisory Group of EU and non-EU states (USA, UK, Germany, Israel, Australia, France and Spain). After the Madrid train bombings, the Greek government requested NATO help in providing AWACS flights, Standing Naval Force Mediterranean patrols, CBRN response and intelligence.

The Athens Olympics also provided an interesting example of public–private partnerships in MES (Migdalovitz, 2004). The Greek government awarded a $c.\$250m$ Olympics security infrastructure contract to a consortium led by the US SAIC corporation. The consortium included Siemens, Nokia, AMS, E Team and Greek companies (ALTEC, Diekat and Pouladis-PC Systems). The contract covered the provision of security measures at sporting venues, the Olympic Village and ports. These measures included building security command centres and a CCTV system. The wider issues of PPPs in MES are considered further in the next section.

The UK 2012 London Olympics preparatory process and planned delivery mode show similar features to Athens 2004, and are therefore not inconsistent with EU guidelines. For example, the lead security body is the 'Cabinet-level Olympics Security Committee, chaired by the Home Secretary and consisting of representatives of UK security and resilience agencies …' (Jennings and Lodge, 2009). Furthermore, the operation of Olympic events is to be managed by the London Organising Committee for the Olympic Games (LOCOG), which is a private company owned by the UK government. Moreover, it has been noted that 'security for the Games entails a complex network of public and private organizations …' (ibid.). Currently, projected figures for security personnel deployments suggest approximate figures of 15,000 police officers and 7,000 PSI personnel.

The potential contribution that the EU might be able to make to the security of the 2012 London Olympics has recently been examined, inter alia, by the House of Lords EU Committee (2009). In particular, the Committee looked at the EU's 'Civil Protection Mechanism', which was established in 2001, and its major incident related Monitoring and Information Centre (MIC). The MIC can act as a channel for arranging mutual aid for a major event. For example, Portugal asked other Member States for CBRN decontamination teams to be on standby during the Euro 2004 championships and, as noted above, Greece made a similar request to the MIC for the Athens Olympics (ibid.: Q32) and France has also requested that the MIC be on standby for major sporting events (ibid.: Q40). However, as of the Report's date (March 2009), the UK had made no approaches to the MIC (ibid.: 13). Indeed the Home Office's 2012 Olympics counter-terrorism strategy summary statement merely refers vaguely to the UK working with, inter alia, 'international partners' (Home Office, n.d.).

By comparing the more detailed expectations that the EU has of an Olympics host Member State with the unclassified version of the Home Office's *London*

Table 11.1

EU 2004 Olympics Handbook requirements	Home Office (2009) Olympics Safety and Security Strategy Provisions
Rolling threat assessment and risk analysis linked to security provision responses	Yes – via OSSSRA (Olympic Safety and Security Strategic Risk Assessment (paras 17, 18 and 19)
Clearly set out duties of all authorities and services	Yes (paras 3 and 4)
Permanent national contact point	The Strategy has a Senior Responsible Owner (SRO) from the Office for Security and Counter Terrorism (OSCT) in the Home Office
Draw up strategic, operational and tactical plans	Yes (paras 45–50)

2012 Olympic and Paralympic Safety and Security Strategy (Home Office, 2009), it is possible to note some congruence, but *not* to infer the influence of the EU in the absence of any mention of EU guidance in the Home Office documents.

Apart from the EU event policing networks, referred to earlier, that can obviously also contribute to any MS holding an ME like the Olympics, Europol has also been providing dedicated assistance. This takes the form of location-specific threat assessments on terrorism and organised crime, an event operational support team at Europol headquarters, and a Europol liaison officer with the relevant police force. Europol has provided this form of support for the 2004 Athens Olympics and the 2006 Turin Winter Olympics (Europol Press Releases, 18 January 2006 and 19 May 2006).

Event security partnerships

As can be seen from the Athens Olympics, discussed above, a Member State will often need to be in some form of partnership with the private sector over event security because of the private ownership of the event venue and the public sector's need for supplementary security resources from the private sector to off-set the opportunity cost of competing demands for public sector policing resources. PPPs in MES in the EU are built upon a number of core understandings and principles. In all MS, the ownership and responsibility for physical space is divided between public authorities (mainly sub-national local government bodies and the police) and private entities, both individual and corporate. In a similar manner, the attendant security responsibilities are so divided, including those applicable to the general public when present on either public or privately owned spaces.

The basic principle underlying PPPs in MES has been explicitly stated by the EU Council in the proposition that the event 'organiser has the primary responsibility for the event ...' (EU, ENFOPOL 119, 2007). It is likely that in most MS, the event organiser's (MEO) responsibilities will be set out in a written format, with some form(s) of licence(s) for a private sector organiser, and that either in these documents or linked documents the police duties will be specified. A common expectation of police duties within an ME 'footprint' is that they will respond to criminal activity, public order situations or security incidents beyond the management capacity of the MES PSI provider. Normally such MES responsibilities are only applicable to the 'event footprint' (e.g. a stadium and its car parks – see, for example, the MEO responsibilities set out in the French law of 1995), but, in the UK, the police are seeking to recover the costs of what they call the 'consequential policing' of the adjacent public spaces to the ME area, although the Home Office states that this is not currently permitted (see House of Commons Home Affairs Committee, 2009, 2010).

This core principle of MEO MES responsibility is equally applicable to both the public and private sectors. It can be linked to a general understanding that, in delivering MES, a distinction can be made between *partners* (defined as those with a direct responsibility for delivering all or some part of an event security plan) and *stakeholders* (defined as those with no direct security responsibilities but who may influence the shape of the security operation). Common examples of partners are football club or conference centre security staff. A monarch or president may be a stakeholder in the sense of, for example, requiring MES provisions to be as unobtrusive as possible and to allow maximum interaction with the general public. In a similar vein, a cultural major event impresario stakeholder may want security to have a 'light touch' so that the 'stars' have easy publicity shots. Whilst these terms may be generally acceptable, it was noted at an EU Seminar on PPPs in crime prevention in 2002 that 'various EU countries had different understandings of the definition of PPP ...', with, for example, Germany seeing the PSI as a partner and the Netherlands seeing the PSI as a contractor.

The PSI as an MES partner

The private sector providers of security services can be divided into four categories. The first category is the private security industry (PSI) as normally understood working within MS which provides trained security personnel for, inter alia, entry controls, crowd control and surveillance, and associated technical security systems. This category of PSI companies is likely to operate as providers of both crime prevention services and MES services (see Jones and Newburn, 2006; Hess, 2009). The UK public sector distinguishes in its relationship with the PSI between situations where PS services are directly employed by an event organisation or event location organisation, and situations where the security provision has been contracted out.

In the second category are those companies that provide armed military-style security outside the EU area, usually in conflict areas or where the operations are in zones within a country where the government's rule is contested. The companies in this category are usually called 'Private Military and Security Companies' (PMSCs), and are sometimes described as 'mercenaries'. PSI companies in EU states can offer armed security services within Member States but under strict national regulation. The PMSCs are not likely to be utilised for MES within the EU.

The third category comprises situations in which the public police forces of MS provide security services within the private space of a ME, like a major football match, but where the MEO meets the costs of this form of police presence. In some MS, this category of PPPs for a range of security duties, including MES, is quite formally institutionalised. In Lithuania, for example, there is legal provision for a 'Public Police Protection Office' (PPPO), although with the development of the PSI in Lithuania the permitted functions of the PPPO have been scaled down by a law of 1998 (Juska, 2009: 235, 237).

The fourth category covers an MEO using private citizens who either volunteer as event stewards or are employed solely in a safety stewarding capacity, a PS category commonly found at some football matches, charity events or cultural events. With regard to this last category, two issues arise: first, the level of training given to or required for volunteer or paid stewards and, second, event security management problems that may arise in MES situations where police, PSI, paid stewards and volunteer stewards are present together at an event location.

The EU, to date, has not specifically addressed the issue of PPPs in MES and has only addressed the issue of PPPs in the context of ensuring that PPPs were rule compliant with Community Law on Public Procurement and Concessions through a Green Paper or consultative process and the issuing of a subsequent Commission Communication (European Commission, 2004, 2005a, b). In that process, it was noted that the 'term public–private partnership ("PPP") is not defined at Community level ...' but that it was generally understood that 'the term refers to forms of cooperation between public authorities and the world of business ...' (European Commission, 2008b). However, the EU Commission does encourage PPPs in MES in order to foster both the security equipment industry and the private security services industry.

In recent decades, as Hakala observes:

> changes in crime patterns and the 'war on terror' have meant a new allocation of police resources and a need to assess whether private security could be utilised more widely to support state-run security ... [but] a prerequisite for expanding the role of private security providers in this context is an improvement of security officers' screening, training, and opportunities to co-operate with authorities ... [and] an example of this can be seen in the security arrangements of big international conferences and sports events today, the security requirements of which are so huge that without private security the police cannot handle them.
>
> (Hakala, 2008: 16–17)

There is a general recognition of the importance of security training for the PSI and this was recently re-emphasised in the ESRIF Report (ESRIF, 2009; see also

Commission of the European Communities, 2009) which endorses both 'transnational initiatives in training and education for security functions ...' and the 'use of virtual reality, "gaming" and other simulation environments ...' (ESRIF, 2009: 17).

Brodeur (2007), reflecting further on his 'high and low policing' typology, also noted that, post-9/11, there are signs of the PSI entering 'high policing' areas (i.e. state interest protection as opposed to private interest protection), although he further endorsed De Waard's 1999 conclusion that the PSI in Europe is 'the secondary source of protection' (Brodeur, 2007: 31). It might be posited, nonetheless, with reference to MES, that in terms of private MEs, the public sector does assume that the PSI will be the initial primary source of protection at the event location. However, in terms of PSI roles, Brodeur proposes the hypothesis that 'the higher the stakes in security the more will the responsibilities of public government be (re)asserted' (ibid.). Brodeur bases this hypothesis on 'the need for *legitimacy* in the provision of security ... [and] the issue of *symbolic* power ...' (ibid.).

During the last three decades or so, several forms of public–private security relations have engendered vigorous public debate (Dorn and Levi, 2007). First, concern has been raised over what may seem to be attempts by governments to cut the costs of providing professional public police services by encouraging certain varieties of private security provision or, indeed, outsourcing some public security duties to the PSI (see Wakefield, 2003; Chesterman and Fishrer, 2010). Second, PS, whether by direct employment or subcontract, has been considered, more positively, to be a useful partner in both crime prevention and anti-social behaviour control (see Jones and Newburn, 2006). Two common examples are those of shopping centre security staff and football club stewards. Third, there are the consequences of 'the privatisation of industries and services belonging to societies' critical infrastructure ... which need a new model of protection' (Hakala, 2008: 3). A common thread linking all these PS activities originating within the EU has been a well-established recognition of the need for all of these activities to be subject to some form of state-law-based regulatory framework.

The domestic regulation of the PSI in EU MS is both variable and dynamic in character. Hakala notes that the majority of EU MS regulate the PSI under the authority of an interior-type ministry, which suggests either the recognition of its quasi 'policing' type role or the desire to clearly distinguish between the functions reserved to the public sector police and those permitted to the PSI (Hakala, 2007). Regarding the nature of the PSI in the EU, Hakala's 2007 study, which included coverage of the EU-25 states, found that the provision of crowd control/event security was the second-ranked most common PSI function (89 per cent).

As Dorn and Levi note, however,

> the terms of the debate on governance are by no means settled.... [I]n the European Union context, there has been considerable controversy over the extent to which private security should be defined primarily as an aspect of and integral to national security or, alternatively as an aspect of commerce.
> (2007: 214)

In a bureaucratic sense, the EU recently made a pillar location choice by moving the public–private security dossier from DG Justice Freedom and Security (Third Pillar) to DG Enterprise and Industry (First Pillar) (Dorn and Levi, 2009: 302).

Dorn and Levi's study of the security industry in 2006–2008 was for a European Commission funded project which sought to explore the options for an 'over-arching' structure for a pan-European public–private dialogue on 'all aspects of terrorism and serious crime' (ibid.). Currently, the main European Commission backed public–private security dialogue that has been established (in 2007) is the European Security Research and Information Forum (ESRIF – see Hayes, 2009). ESRIF has so broad an issue coverage that it has only a limited direct connection with MES issues. ESRIF's 2009 report defines a European Security Research and Innovation Agenda (ESRIA) which might have some relevance to MES in terms of major incident prevention and response, but ESRIF does not cover matters to do with the actual practice of MES. An examination of the full ESRIF 2009 Report has found a small number of possible MES-related points, such as the use of MES incident case studies, common security training, and a proposal for a European 'quality' mark for security products and services.

The European Commission has also specifically expressed an aspiration to explore the feasibility of creating a 'pool' of EU MES equipment, and this might be linked to the Commission's goal of opening up the market in defence and security procurement. This goal has been given a recent impetus by the entering into force (21 August 2009) of the Directive on defence and security procurement (Directive 2009/81/EC), although MS will still be able to use A296(EC) to exempt certain defence and security procurement projects from EU public procurement law on national security grounds (EU Press Release, August 2009).

Conclusions

The EU's involvement in matters relating to MES stems from the EU's capacity, with the support of MS, to address matters of common concern. However, that capacity is constrained by: EU treaty provisions, the relatively small EU budget and EU institutional capacity limitations. It is further constrained by the fact that domestic policing remains the sole preserve of the MS. In EU terms, this point has been reflected in the way that matters relating to policing and security cooperation have long been treated as an EU intergovernmental policy area, as opposed to a policy area subject to the EU's supranational law-making powers, though this situation has been somewhat modified by the Lisbon Treaty.

The dominant EU mode of response to matters of common concern over MES issues has been via the Council of Ministers, with national expert advice, drawing up manuals of guidance on MES. These may be said to represent EU-level agreements on standard setting or 'best practice'. However, they only have the force of guidance documents for MS. The European Commission has used its various policy initiative powers to fund, through Framework Programme 7, the long-term EU–SEC project on MES, which utilises the MES experience of participating Member States' police and interior ministry officials to draw up further forms of MES guidance and to propose, inter alia, the sharing of research on MES. This initiative also links in with the Commission's aspiration to promote the internal security equipment industries of the MS through the ESRIF process. The Commission further seeks to promote the development of the EU market for PSI services, as the services of PSI companies are clearly an integral part of MES in MS, to varying degrees.

Olympics MES policy is just a sub-set of EU MES policy, although of course a high profile one. In summary, the EU provides for MES, including the Olympics when held in MS, guidance manuals on MES, and MS can make use of the services and expertise of Europol and the EU civil protection bodies. However, it should be noted that, for MES security threat analysis, Europol is heavily dependent on intelligence inputs from MS. In practice, an MS, which is an Olympics host state, is free to utilise both inter-governmental help, both from EU and non-EU states, and help from EU sources. The Greek case illustrates this point very well. The degree to which an EU Olympics host MS utilises these forms of external help will be a variable factor related to national MES capacity. At a minimum level, an EU Olympics host state is likely to provide an Olympics MES system which conforms to the broad scheme contained in the EU Olympics' manual and to be open to relevant external sources of help as a way of demonstrating national 'due diligence'. This practice has been illustrated, earlier, by the references to the provisions of the UK's *London 2012 Olympic and Paralympic Safety and Security Strategy* (Home Office, 2009).

Author's note

The author of this chapter has been involved with the EU–SEC Project since 2004 as the academic member of the UK Team (the UK Team has its lead professional members provided from the Metropolitan Police) but the views expressed in this chapter are solely those of the author.

Note

1 Five EU States – France, Italy, the Netherlands, Portugal, Spain – have formed, by inter-governmental initiative, the European Gendarmerie Force and declared elements of their gendarmerie forces available to the EGF for EU CIVPOL missions outside the EU area. Recently, Romania's Gendarmerie has been admitted as the sixth member of the EGF; see www.eurogendfor.org (accessed 3 February 2010).

Bibliography

Bache, I. and George, S. (2006) *Politics in the European Union*, Oxford, Oxford University Press, 2006.

Body-Gendrot, S. (2003) 'Cities, security and visitors: Managing major events in France', in Hoffman, L.M., Feinstein, S.S. and Judd, D.R. (eds) *Cities & Visitors: Regulating People, Markets and City Spaces*, Oxford, Blackwell, p. 40.

Brodeur, J.-P. (2007) 'High and low policing in post-9/11 times', *Policing*, 1, 1.
Chesterman, S. and Fisher, A. (2010) *Private Security and Public Order*, Oxford, Oxford University Press.
CoESS (2004) *Panoramic Overview of Private Security Industry in the 25 Member States of the European Union*.
Commission of the European Communities (2009) *A European Security Research and Innovation Agenda – Commission's Initial Position on ESRIF's Key Findings and Recommendations*, COM (2009)691 final, Brussels 21 December.
Council of the EU (2001) *Special Units to Guarantee the Safety of Meetings of the European Council and Other Comparable Events* (Doc. 11934/01), ENFOPOL 96, Brussels, 20 September.
Council of the EU (2002) Security Handbook for the Use of Police Authorities and Services at International Events Such as Meetings of the European Council (Doc.12637/3/02), REV 3 LIMITE ENFOPOL 123, Brussels, 12 November.
Council of the EU (2004a) *Handbook for Member States Co-operation Against Terrorist Acts at the Olympic Games and Other Comparable Sporting Events* (Doc. 5744/1/04), REV 1 LIMITE ENFOPOL 14.
Council of the EU (2004b) *Proposals Relating to the Enhancement of Measures to Combat Football Related Violence* [Doc. 7017/04], ENFOPOL 23 (Note from the Presidency to the Police Cooperation Working Party). The latest version of the 'Football Handbook' is ENFOPOL 159, EU Council 13119/06, LIMITE, Brussels, 16 October 2006.
Crawford, A. (ed.) (2002) *Crime and Insecurity: the Governance of Safety in Europe*. Devon, Willan Publishing.
Dorn, N. and Levi, M. (2007) 'European private security, corporate investigation and military services: collective security, market regulation and structuring the public sphere', *Policing and Society*, 17, 3.
Dorn, N. and Levi, M. (2009) 'Private–public or public–private? Strategic dialogue on serious crime and terrorism in the EU', *Security Journal*, 22.
Edwards, A. and Hughes, G. (2005) 'Comparing the governance of public safety in Europe: a geohistorical approach', *Theoretical Criminology*, 9, 3, 345–363.
ESRIF (2009) *ESRIF Final Report*, December.
EU Council (2007) *Handbook for Police and Security Authorities Concerning Cooperation at Major Events with an International Dimension* (EU Doc. 10589/1/07), REV 1, ENFOPOL 119, Brussels, 4 July.
EU Press Release (2006) Green Paper on Detection and Associated Technologies in the Work of Law Enforcement, Customs and Other Security Authorities, COM(2006) 474 final, Brussels, 1 September.
Europa (2006) *Coordinating National Security During Major Events in Europe*, Press Release IP/06/1699, 7 December.
Europa (2009) New Directive on Defence and Security Procurement Enters into Force, Press Release IP/09/1250, Brussels, 25 August.
European Commission (2004) *Green Paper on Public–Private Partnerships and Community Law on Public Contracts and Concessions*, COM(2004)327.
European Commission (2005a) *Communication on Public–Private Partnerships on Community Law on Public Contracts and Concessions*, COM(2005)569.
European Commission (2005b) *Report on the Public Consultation on Public–Private Partnerships and Community Law on Public Contracts and Concessions*, SEC(2005)629.

European Commission (2008a) *Coordinating National Research on Security During Major Events in Europe*.
European Commission (2008b) *Initiative on Public–Private Partnerships and Community Law on Public Procurement and Concessions*, www.ec.europa.eu/internal_market/publicprocurement/ppp_en.htm (accessed 13 November 2009).
European Commission (2007) *Commission Promotes Public–Private Dialogue to Improve Security of EU Citizens*, Press Release IP/07/1296 (on ESRIF)11 September.
Europol Press Releases (2006)
Guild, E. and Geyer, F. (2008) *Security versus Justice?* London, Ashgate.
Gruszczak, A. (2010) 'Internal security strategy for the European Union', *Statewatch Briefing*, January.
Hakala, J. (2007) *The Regulation of Manned Private Security: a Transitional Survey of Structure and Focus*, Department of Sociology, School of Social Sciences, City University, 18 March.
Hakala, J. (2008) *Why We Regulate Private Security?* CoESS Report, Helsinki.
Hayes, B. (2009) *NeoConOpticon – The EU Security–Industrial Complex*, Transnational Institute & Statewatch, London.
Hess, K.M. (2009) *Introduction to Private Security*, Wadsworth, Cengage Learning, 4th edn.
Home Office (n.d.) *Counter-Terrorism Strategy – Securing the 2012 Olympic and Paralympic Games*, http://security.homeoffice.gov.uk/counter-terrorism-strategy/olympics2012/ (accessed 18 February 2010).
Home Office (2009) *London 2012 Olympic and Paralympic Safety and Security Strategy*. London, Home Office.
House of Commons Home Affairs Committee (2009) 10th Report of Session 2008–09, 'The Cost of Policing Football Matches', HC 676, July, London, Stationery Office.
House of Commons Home Affairs Committee (2010) 2nd Special Report of Session 2009–10, 'The Cost of Policing Football Matches: The Government's Response to the Committee's Tenth Report of Session 2008–09', HC 339, February, London, Stationery Office.
House of Lords EU Committee (2004) 20th Report of Session 2003–04, 'Security at EU Council Meetings', HL 119, London, Stationery Office.
House of Lords EU Committee (2009) 6th Report of Session 2008–09, 'Civil Protection and Crisis Management in the European Union', HL 43, March, London, Stationery Office.
Jennings, W. and Lodge, M. (2009) *Governing Mega-Events: Tools of Security Risk Management for the London 2012 Olympics and FIFA 2006 World Cup in Germany*, Paper for the 59th Political Studies Association Conference, Manchester, 8 April.
Juska, A. (2009) 'Privatisation of state security and policing in Lithuania', *Policing and Society*, 19, 3.
Jones, T. and Newburn, T. (eds) (2006) *Plural Policing: a Comparative Perspective*, Oxford, Routledge.
Migdalovitz, C. (2004) *Greece: Threat of Terrorism and Security at the Olympics*, CRS Report for Congress, RS21833, 30 April.
Ministry of Justice (the Netherlands) (2003) Report of the Seminar 'Public–Private Partnership in Crime Prevention', held in The Hague, 16–17 December 2002.
Monar, J., Rees, W. and Mitsilegas, V. (2003) *The EU and Internal Security*, London, Palgrave Macmillan.
Ocqueateau, F. (2006) 'France', in Jones, T. and Newburn, T. (eds) *Plural Policing*, London, Routledge, p. 73

Presidency of EU Council to Police Cooperation Working Party (2006) *Proposal for Security Handbook for the Use of Police Authorities and Services At International Events* (EU Council doc. 15226/1/06) REV 1, ENFOPOL 190, 22 November, Para I.2.

Wakefield, A. (2003) *Selling Security: the Private Policing of Public Space*, Cullompton, Willan.

Walker, N. (ed.) (2004) *Europe's Area of Freedom, Security and Justice*, Oxford, Oxford University Press.

Winfield, G. (2006) 'Population explosion', *NBC International*, Autumn.

12 Critical reflections on securing the Olympics
Conclusions and ways forward

Pete Fussey, Anthony Richards and Andrew Silke

This book has aimed to explore a number of critical themes around the challenge of delivering a safe and secure Olympics in the context of the serious contemporary and 'enduring' international terrorist threat. The first section outlined the nature of this threat to the UK and what the potential threats might be to the 2012 Olympics in particular, though many of the insights will also apply to future events. The second section, beginning with a useful contextual discussion of risks in general to the Olympics, has discussed the key 'responding' themes of transport security, surveillance and designing stadia for safer events; and the final section has discussed the role of, and linkages between, institutions: including the challenge of achieving effective multi-agency coordination, the role of the private security industry and European perspectives on major event security.

It has not been part of the remit of this volume (and the first section in particular) to outline what the causes of contemporary terrorism might be,[1] nor has it sought to provide an overview as to what a robust and comprehensive counter-terrorism strategy in general might look like. Rather, it has outlined the contemporary threat of terrorism and what the implications of this might be for the Olympics – with a particular eye on 2012 – as well as exploring some response themes that are especially pertinent to the Olympics.

Nevertheless, in any discussion as to how to deliver a 'safe and secure' Games, some consideration as to how to mitigate the threat from terrorism more broadly is warranted. In the editors' view, any response to terrorism must be informed by an understanding as to what has caused it in the first place. This may seem obvious, but it has been apparent from some of the more draconian proponents of the 'war on terror', or indeed from some of those who would deny the impact of Iraq and Afghanistan on international terrorist recruitment, that an understanding as to *why* terrorism takes place is not the priority. Rather, the emphasis for some is to 'defeat the enemy' or to 'eliminate the terrorists' wherever they may be, and whatever the longer-term consequences might be.

The UK and other countries have both been introducing and considering some unprecedented technological solutions (and, indeed, this is an important part of counter-terrorism), but it should never take precedence over an honest appraisal as to what is motivating young males in particular into sacrificing themselves in the first place, or what it is that makes them more susceptible to terrorist recruitment.

There is no avoiding this key challenge that must lie at the heart of a well-informed counter-terrorist strategy.

Particularly worrying for the British government, and for London 2012, has been MI5's warning of the significant numbers within the UK's shores who are willing to perpetrate or facilitate mass casualty attacks. It suggests that many are willing to buy into Al Qaeda's ideology and its message of bringing about change through violence. Bin Laden's doctrine and interpretation of the conflicts in Kashmir, the Middle East, Iraq and Afghanistan, disseminated globally via the mass media and the Internet, represents the ideological gel that underpins the threat in the UK and elsewhere. Despite the links that have been identified between 'home-grown' terrorists and extremists abroad (in Pakistan, in particular), the threat is essentially a decentralised one and it is this that makes Al Qaeda's ideology and narrative so important as a binding influence.

Arguably, therefore, one of the most important challenges is to counter the narrative of Al Qaeda and its portrayal of a conflict between Islam and the West. This particular priority has been outlined in the British government's latest *Strategy for Countering International Terrorism*. In the section entitled 'Counter-terrorism Communications' the document acknowledges the imperative need to counter and disrupt this narrative by outlining the purpose of the new Research, Information and Communications Unit (formed in June 2007) that includes 'exposing the weaknesses of violent extremist ideologies and brands' and 'supporting credible alternatives to violent extremism using communications' (UK Home Office, 2009: 153).

Emphasising the 'demotivation' part of terrorism is very much what the 'Prevent' strand of the UK's counter-terrorism strategy, CONTEST, has sought to do. It is, of course, not a complete response to terrorism – however successful states might be in reducing the propaganda and recruitment opportunities (and therefore the motivation) for terrorism, there will always be a hardcore who persist. Northern Ireland provides a good example of this. The three governments of the UK, the Irish Republic and the United States have expended enormous political will and effort into bringing about and nurturing the peace process that has led to greater demotivation for terrorism. Yet, not only has the threat from Dissident Republicanism remained, but it has markedly increased in recent years. There is also, therefore, a need to 'pursue', to 'protect' and to 'prepare' against terrorism.

The UK has been credited with having a comprehensive counter-terrorism framework through its CONTEST strategy, outlined first in its July 2006 document, *Countering International Terrorism: the United Kingdom's Strategy* and then updated (as CONTEST II) in March 2009 as *Pursue Prevent Protect Prepare, The United Kingdom's Strategy for Countering International Terrorism*. It has revamped its civil-contingencies legislation in dealing with major disasters and terrorist events, MI5 has expanded to nearly 4,000 personnel, and additional resources have enhanced the UK's emergency response preparedness (through, for example, the Fire and Rescue Service's New Dimensions Project).

One persistent criticism of the CONTEST strategy is that it has become too heavily focused on Islamist extremism and has lost sight of many of the other threats still facing the UK. The already-mentioned dissident Irish Republican groups, for example, have carried out far more terrorist attacks in the UK in the past ten years than the Islamists, but there is no serious discussion in security circles about counter-radicalisation or de-radicalisation strategies in terms of dissident Republicans. There are currently just over 100 prisoners associated with Al Qaeda in the UK prison system who were convicted of terrorism-related offences, and serious efforts have been underway for some years to develop prison-based programmes to help de-radicalise these prisoners. In contrast, there are some 80 Irish paramilitary prisoners also in UK jails and no efforts whatsoever have been made to develop a de-radicalisation programme aimed at them.

We have to be careful of the Al Qaeda tunnel-vision effect – where all other groups and threats tend to get lost in the background. Al Qaeda are not the only threat to the Olympics. Indeed, while Al Qaeda has shown a strong interest in hitting past Olympics, the organisation has to date never actually been able to mount a successful attack against the Games. This is in contrast to a range of other groups who have been able to carry out attacks. While Al Qaeda are one of the few groups who would be willing to deliberately kill very large numbers of people at such an event, others will also see benefits in targeting the Games in other ways.

There are other potential problems that can also have direct implications for the Olympics should a terrorist attack take place during Games time. For example, nearly four years after the events of September 2001, and in the expectation that it was 'when' rather than 'if' the UK would suffer from a major terrorist attack, there appeared to be some serious shortfalls in the response to the July 2005 attacks, including unreliable communications between the emergency services (and even within them), especially in underground environments, and a lack of planning and care for survivors (Richards, 2007; Weston, this volume).

Nevertheless, the post-9/11 environment – and subsequently the post-7/7 environment – has seen major developments in UK counter-terrorism that have provided the underpinning context for the security of the Olympics in 2012. Therefore, rather than seeing the Olympics as a one-off security challenge that requires separate attention and a whole new security apparatus, any specific Olympic-related measures in the UK have instead been viewed as add-ons or as providing 'additionality' to what is already in place. At the national level, the impetus for any 'additionality' for Olympic security is evident through the existence of the Olympic Security Directorate in OSCT and through the *London 2012, Olympic and Paralympic Safety and Security Strategy* (2009a). Although 'prevent' and 'pursue' are also integral parts of counter-terrorism strategy, and therefore to the security of the Olympic Games, the document, as Weston in this volume explains, outlines its aims as to protect, to prepare and more particularly for the Olympics, to *identify and disrupt* threats to the safety and security of the Games; *command, control, plan and resource* the safety and security operation; and *engage* with international and domestic partners and communities to enhance our security and ensure the success of our Strategy (OSCT, 2009a). Table 12.1 shows the framework OSCT has used to organise and structure the different projects.

The OSCT also states that the strategy will be delivered in relation to time and place, time being split into the four phases of: (1) design, plan and build; (2) overlay and testing; (3) Games time; (4) recovery and decommissioning; and place into (1) competition venues; (2) non-competition venues; (3) training venues; (4) training camps; (5) transport networks, (6) Olympic route network (designated routes, i.e. between venues and so on) (OSCT, 2009a). It also outlines the role of the Olympic Safety and Security Strategic Risk Assessment (OSSSRA), also noted by Weston, which 'will evolve as understanding of risks unfolds and will inform strategic level decision-making and the prioritisation of resources', and the 'Olympic Intelligence Centre' which will regularly update a Strategic Threat Assessment by bringing together 'assessments by existing agencies with responsibility for collecting intelligence,' and the Olympic Security Board which provides general oversight (OSCT, 2009a).

Thus, Olympic security is being planned in the context of the recent and major overhaul in civil contingencies and counter-terrorism capabilities the latter of which in particular has been prompted by 9/11 and the international terrorist threat environment. Even though the Olympics are rapidly approaching (at the time of writing), threat assessments might change and/or different terrorist threats may emerge. For example, dissident Republicans are currently restricting their campaign of violence to Northern Ireland, but Republican movements in the past have often seen great value in conducting attacks in England, and particularly in London (McGladdery, 2006). As a result, Olympic security can never be static (a fact explicitly acknowledged in the Olympic Security Strategy) but has to evolve and adapt to the prevailing threat environment. International events can have a direct bearing on domestic security and so events abroad in the run up to the Games can also have serious implications for Olympic security.

Reflections on terrorism and Olympic security

This volume has demonstrated that, given its exceptional and unprecedented scale, Olympic security strategies require a blend of many components, each involving variations in efficacy when dealing with terrorist violence. In addition, such large-scale security operations raise a number of key questions regarding which threats are being mitigated, what is being secured, where, and for which period? These issues are now discussed in turn.

Disparate risks

Since Munich, the Olympics have been threatened by the full complement of different terrorist groups, including left-wing radicals (Barcelona, Nagano, Athens), left-wing state proxies (Seoul), Christian fundamentalists (Atlanta), ethno-nationalist separatists (Sarajevo, Los Angeles, Barcelona, Beijing), single-issue

Table 12.1 OSCT work stream structure for Olympic safety and security projects

Protect	Prepare	Identify and disrupt	Command, control, plan and resource	Engage
People	Olympic resilience	Olympic intelligence	Resources	International relationships
ID assurance	Specialist response	Serious and organised crime	National resource requirement	Community relationships
VIP protection	Critical Olympic supporting infrastructure	Volume crime	Meeting demand	Volunteers
Venue		CCTV	Training	Prevent
Site and venue security		Automatic number plate recognition	Operations logistics infrastructure	Industry
Chemical, biological, radiological, nuclear and explosive			Command and control	
Non-venue			National co-ordination	
Transport security			Airwave	
Border security			Olympic control infrastructure	

groups (Albertville), hostile states (Seoul), as well as violent Jihadi extremists (Sydney). Pete Fussey provides a list of the different terrorist attacks against the Olympics at the end of this volume which starkly illustrates the surprising range of groups that have been involved. Although the Olympics may have symbolic utility for a variety of terrorist groups with often-distinct targeting strategies, components of Olympic anti-terrorist strategies may achieve more success when tackling some groups than for others (e.g. Fussey, 2007, for the case of CCTV).

Of course, it is crucial to retain perspective as Olympic risks extend far beyond terrorism. Alongside increases in more routine crimes within host communities (Decker *et al.*, 2007), the Olympics may also provide the backdrop for more serious crime, some perhaps involving particularly grievous forms of victimisation. For example, as the experience of the Atlanta, Sydney and Athens Summer Olympiads articulates, the sex industry has become an integral component of the Olympic marketplace (Eleftherotypia, 2003; Graycar and Tailby, 2000; Hellenic Republic Ministry of Public Order, 2004). With London's sex industry already holding intrinsic connections to human trafficking and other forms of severe exploitation (EAVES, 2008), these considerations are serious and pressing. Indeed, Fussey and Rawlinson's (2009) ongoing ethnographic research into organised crime in East London has yielded data pointing to the adaptability of the sex industry in two of the capital's Olympic Boroughs. Elements include the increased mobility (and reduced detectability) of the industry; transferability of skills as established organised criminal elements have been entering new market positions; and the establishment of new trans-ethnic coalitions as nodal points that facilitate the industry. Most (Summer and Winter) Olympics are hosted in suburban theme parks away from core urban settings (with the possible exception of Atlanta). Given that the London Olympics has and will be an employer on a massive scale, and that it will host nine million ticket holders in a thriving urban area with an established informal economy, such risks of severe victimisation constitute a serious issue.

Temporal and spatial threats

Returning to the central theme of terrorism and the Olympics, despite the inevitable spotlight centred on the events at Munich and Atlanta, historically, terrorist activity has had an increased scale and intensity during the run-up to the Games. Such incidences are evinced by the terrorist activities preceding the Seoul, Barcelona and Athens Games. For example, to discourage visitors to the Seoul Olympics, North Korean agents and Japanese Red Army proxies launched multiple attacks against South Korea's commercial aviation industry with the cost of 115 lives between November 1987 and May 1988 (United Nations Security Council, 1988). Between the IOC's award of the Games to Barcelona in 1987 and the opening ceremony in 1992, the left-wing Grupo de Resistencia Antifascista Primo Octobre (GRAPO), Catalan separatists Terra Lliure and Basque separatists ETA all conducted attacks aimed at disrupting the Olympics or inflicting reputational harm. In Greece, the left-wing Revolutionary Struggle group used five bomb attacks in the lead-up to the 2004 Games to lever greater exposure for their cause. All of these attacks also took place away from Olympic sites and connect with the more enduring themes surrounding displacement (explored in Silke, this volume). In addition, terrorist threats and perceptions of them may also become elevated *after* Olympic attacks. Buntin (2000), for example, notes how, following Eric Rudolph's bomb in Centennial Park, the authorities were called to attend suspicious packages for an average of every ten minutes until the end of the Games. Together, such events demonstrate that significant terrorist threats are far from confined to the temporal and spatial boundaries of the Olympic spectacle and, moreover, raise considerable implications for security planning and resourcing.

One potential means of conceptualising the collection of security measures deployed to mitigate such wide-ranging and disparate risks is in terms of five concentric 'Olympic rings of security'.[2] Comprising an amalgamation of human and technical strategies, these can be viewed as operating within different spatial and operational contexts. Building on the many themes raised in this volume, these five rings can be defined as follows:

Ring 1. National Boundaries: including aviation and port security, traditional security provisions, passenger screening, behavioural analysis, detection technologies, explosives sensors.

Ring 2. Locality: the host city and proximate region – including local and national policing arrangements, extant security infrastructures (i.e. CCTV), new security coalitions including the Olympic Security Directorate and networks incorporating private security actors and volunteers.

Ring 3. Transport: drawing on policing bodies such as the British Transport Police, existing transport security technologies (i.e. ANPR), additional monitoring of major routes leading towards the Olympic 'area' (i.e. M11 and A12), securitised transport corridors, traffic-free 'buffer zones' and the implementation of new technologies (i.e. the embedding of RFID technologies into event tickets).

Ring 4. Venue: security of the Olympic Park (or 'Island Site') – including perimeter security (currently protected via electrified fence), secure by design and defensible space motifs (see Coaffee, this volume), surveillance strategies, potential deployment of biometric technology, remote sensing, reinforced communication infrastructures and access controls.

Ring 5. Event: including (in addition to the above features), crowd management, enhanced voluntary provision, personal protection and secure stadia design.

Across time and place: 2012 security legacies

Although many of the above security provisions are over-layered onto existing processes and infrastructures, the atypical nature of Olympic security raises a number of further issues and tensions. For 2012, this shift from routine to exceptional security is underpinned by temporary and novel (vertical and lateral) organisational structures, new extended coalitions with the broader public, private and voluntary 'policing family', and the introduction of new technologies of control. While these raise key issues surrounding inevitable difficulties of effective communication and coordination (such as those that plagued the counter-terrorism strategies at Atlanta and Salt Lake City), a further debate concerns the issue of legacy. Although the word 'legacy' is ever-present within 2012 Olympic discourse, it is largely used to denote a very generalised perception of post-Games social and economic improvement. However, in the context of security, there are a number of additional and potentially more complex aspects associated with legacy. For other sporting mega-events in Europe, such as the 2006 FIFA World Cup in Germany and the 2008 UEFA European Championships in Austria and Switzerland, one major security legacy was the continuation of networks of security professionals (Baash, 2008; Hagemann, 2008). For London 2012, important debates remain unresolved over how such security infrastructures will be reconstituted and the manner in which they will reverberate in the 'legacy' period following the Games.

For the immediate locality, tenders for private Olympic contracts have been explicitly encouraged to build 'security legacies' into their bids, thus bestowing the site with substantial mechanisms and technologies of control that may or may not be appropriate or necessary following the Games. Nevertheless, whilst the security priorities of a high profile international sporting event attended by millions of people are rather different from those involved in policing a large urban parkland (the future incarnation for most of the Olympic site), this post-event inheritance of security infrastructures is a common Olympic legacy. Examples of this include the legacy of private policing following the Tokyo (1964) and Seoul (1988) Olympiads (Lee, 2004), and the continuation of zero tolerance style exclusion laws after the Sydney (2000) Games (Lenskyj, 2002). In addition to issues surrounding legitimacy and citizenship, the post-event retention of such measures may impact on the (poorly understood) grievances and experiences of those within East London's 'suspect' communities deemed susceptible to 'radicalisation'. Indeed, London's recent history has demonstrated how the over-policing of particular demographics can generate particular tensions, outcomes and lessons that are difficult to ignore (Scarman, 1981; Macpherson, 1999).

Other social aspects of the 'security legacy' are also important. Another component of this legacy concerns the much-lauded employment opportunities on offer. Of particular relevance is the post-event destination of the thousands of individuals in private security roles. Currently, 7,000 private sector agents are needed to secure the Games effectively, yet leaders of this industry estimate that, at current levels, collectively, only 1,000 individuals could be provided. In response, partnerships with Further Education colleges have been established to divert students towards acquiring the skills needed to match the ephemeral security needs of the Games and thus meet this shortfall. The post-event market of meaningful labour for those possessing such skills, however, is less clear (particularly in a time of economic uncertainty).

Although there are no definitive answers to these issues, in the context of the London Games, security and regeneration issues are, to a degree, linked. As such, endowing post-event security policies with regenerative considerations will not only foment a genuinely inclusive legacy, it will also enhance the provision of secure and safe communities.

Other security legacy matters extend beyond East London and 2012. Mega-event security planning is increasingly becoming standardised and globalised. At a national level, Glasgow will host the 2014 Commonwealth Games, whilst England will host the 2015 Rugby World Cup and the 2019 ICC Cricket World Cup, and may have a chance of staging the 2018 FIFA World Cup. What is clear is that major international sporting events will continue to be hosted by the UK on a reasonably regular basis. Despite their differing configurations, the lessons learnt from the 2012 security operation will have clear relevance to these events. Internationally, the trend towards the standardisation of mega-event security, the mechanisms of ensuring knowledge transfer and the award of Olympic-size events to global cities of a similar stature to London, represents a global relevance of the practical and theoretical issues developed in this book.

Concluding thoughts

Terrorism is a very old problem, which has been around in one form or another for thousands of years. That fact alone should give pause for thought for anyone contemplating an end to terrorist threats, whether against the Olympics or any other set of targets. This is not to say, however, that nothing changes with regard to the security and vulnerabilities of the Games. The protection afforded recent Games has far exceeded that offered to most of their predecessors. The most devastating terrorist attack against the Olympics – Munich – occurred in the context of a security regime that was unbelievably lax compared to what has become standard today.

A recurring theme in different parts of this book has been that absolute security surrounding the Games – or indeed any other major sporting event – is an impossibility. An arms race exists between the attackers and the attacked. Total security is unfeasible – and, arguably, we would not want it even if it could be achieved. In the end, some degree of risk always remains. Nevertheless, proper risk assessment can go a long way in reducing vulnerabilities and improving tactics and strategies for dealing with threats when they do appear. In preparing for Munich, the West German authorities commissioned a security review to examine potential security scenarios the Games might be threatened with. In particular, 26 scenarios were developed and highlighted (Wolff, 2002). Scenario 21 suggested that a group of Palestinian terrorists would penetrate the Olympic village and try to take members of the Israeli team hostage. The scenario proposed the event would start during darkness and that the terrorists would kill one or two hostages at the beginning to demonstrate their seriousness. The terrorists would then make a series of demands (one of which would be the release of Palestinian prisoners in Israeli jails). The scenario predicted that the terrorists would not be willing to surrender the hostages if these demands were refused. The authorities rejected this scenario, informing the creator that they wanted something more realistic.[3]

For modern Games, security planners routinely consider 800 or more potential scenarios when considering threats to the Olympics. One cannot protect against everything, however. Budgets and manpower are still limited. The security budget for Munich was approximately $2 million (or about $12 million in 2008 prices). In comparison, the security budget for the 2012 Olympics is currently estimated at around $1.384 billion, with a serious potential to rise much closer to $2 billion. Even the smaller figure represents more than a hundred-fold increase in spending on security in real terms. This is a huge sum of money, but gaps inevitably remain. Part of the challenge is to understand that vulnerabilities will always exist in some shape or form, and to learn to work with these, rather than ignore them and see only what we want to see.

In the end, the threat of terrorism is a harsh feature of modern life and is a risk that society has had to learn to live with and adapt to. We need to remember, too, that terrorism is not the only risk – and indeed, arguably, it is not even the most serious. Although the explicit theme of this book has been terrorism and the Olympics, many of the chapters have also rightly highlighted that the Olympics and other major sporting events face other critical security challenges, ranging from public order disruption to serious organised crime activity. Some of these challenges are much more prevalent than others.

Sport can be a powerful force for good in human life. The Olympics, in particular, consciously tries to spread and defend many deeply positive messages. We move through ferocious times, however, and not everyone looks at sporting events with benign eyes. For a few, they are an opportunity for mayhem and destruction. Those charged with protecting sporting events face many serious challenges and carry a grave responsibility for attempting to safeguard not only the lives of all those involved, but also with preserving an aspect of the human experience that is truly special.

Notes

1 For this, see, for example, Bjorgo, 2005; Crenshaw, 1981; or, for the contemporary threat in the UK, see O'Duffy, 2008.
2 The authors are grateful to Keith Weston for his contribution to this concept, developed during personal communication.
3 One of the other scenarios involved an extremist group hiring a passenger jet and then crashing the plane into the Olympic Stadium during the Games. This too was dismissed as far-fetched, but is certainly taken very seriously today.

References

Bjorgo, T. (ed.) (2005). *Root Causes of Terrorism*. Oxford: Routledge.

Buntin, J. (2000). *Security Preparations for the 1996 Centennial Olympic Games (B) Seeking a Structural Fix*. Cambridge: Kennedy School of Government, Harvard.

Baash, S. (2008). 'FIFA Soccer World Cup 2006 – Event Driven Security Policies', paper presented at *Surveillance and Security at Mega Sport Events: From Beijing 2008 to London 2012*, Durham University, 25 April.

Crenshaw, M. (1981). 'The Causes of Terrorism', *Comparative Politics*, 13(4), 379–399.

Decker, S., Varano, S. and Greene, J. (2007). 'Routine Crime in Exceptional Times: the Impact of the 2002 Winter Olympics on Citizen Demand for Police Services', *Journal of Criminal Justice*, 35(1), 89–103.

EAVES (2008). *Big Brothel: a Survey of the Off Street Sex Industry in London*. London: EAVES.

Eleftherotypia (2003). 'Olympic Prostitution in Athens', *Eleftherotypia*, 22 June.

Fussey, P. (2007). 'Observing Potentiality in the Global City: Surveillance and Counter-terrorism in London', *International Criminal Justice Review*, 17(3), 171–192.

Fussey, P. and Rawlinson, P. (2009). 'Winners and Losers: Post Communist Populations and Organised Crime in East London's Olympic Regeneration Game', presented at the *2nd Annual European Organised Crime Conference*, Liverpool, UK, 9 March.

Graycar, A. and Tailby, R. (2000). *People Smuggling: National Security Implications*. Canberra: Australian Institute of Criminology.

Hagemann, A. (2008). 'From Stadium to "Fanzone": the Architecture of Control', paper presented at *Surveillance and Security at Mega Sport Events: From Beijing 2008 to London 2012*, Durham University, 25 April.

Hellenic Republic Ministry of Public Order (2004). *Annual Report on Organised Crime in Greece for the Year 2004*. Athens: Hellenic Republic Ministry of Public Order.

Lee, C. (2004). 'Accounting for Rapid Growth of Private Policing in South Korea', *Journal of Criminal Justice*, 32, 113–122.

Lenskyj, H. (2002). *The Best Olympics Ever? Social Impacts of Sydney 2000*. Albany: State University of New York Press.

McGladdery, G. (2006). *The Provisional IRA in England: the Bombing Campaign 1973–1997*. Dublin: Irish Academic Press.

Macpherson, W. (1999). *The Stephen Lawrence Inquiry: Report of an Inquiry by Sir William Macpherson of Cluny*. London: HMSO.

O'Duffy, B. (2008). 'Radical Atmosphere: Explaining Jihadist Radicalization in the UK', *Political Science and Politics*, 41: 37–42.

Richards, A. (2007). 'The Emergency Response: Progress and Problems', in Wilkinson, P. (ed.) *Homeland Security in the UK*. Abingdon: Routledge.

Scarman, L. (1981). *The Scarman Report: the Brixton Disorders, 10–12 April 1981*. London: HMSO.

United Nations Security Council (1988). *Provisional Verbatim Record of the Two Thousand Seven Hundred And Ninety-First Meeting*, 16 February, available at: www.undemocracy.com/S-PV.2791.pdf (retrieved 3 February 2010).

Wolff, A. (2002). 'When the Terror Began', *Time*, 25 August, available at: www.time.com/time/europe/magazine/2002/0902/munich/index.html.

Web references

Office for Security and Counter-Terrorism (2009a). *London 2012, Olympic and Paralympic Safety and Security Strategy*, July, available at: http://security.homeoffice.gov.uk/news-publications/publication-search/general/Olympic-Safety-and-Security12835.pdf?view=Binary.

Office for Security and Counter-Terrorism (2009b). *London 2012, A Safe and Secure Games For All*, July, available at: http://security.homeoffice.gov.uk/news-publications/publication-search/general/Olympic-security-public-sum12835.pdf?view=Binary.

UK Home Office (2006). *Countering International Terrorism: The United Kingdom's Strategy*, July, available at: http://security.homeoffice.gov.uk/news-publications/publication-search/general/Contest-Strategy.html.

UK Home Office (2009). *Pursue Prevent Protect Prepare, The United Kingdom's Strategy for Countering International Terrorism*, HM Government, March, p. 153, available at: http://merln.ndu.edu/whitepapers/UnitedKingdom2009.pdf.

13 Terrorist threats to the Olympics, 1972–2016

Pete Fussey

The chart below illustrates some of the key risks and security challenges that have faced Olympic planners since Munich, 1972. It represents the synopsis of a more expansive and detailed investigation into Olympic threats presented in Fussey *et al.* (2010).

Debates inevitably occur over which activities are 'terrorist' and to what extent they are 'Olympic-related'. This picture is further complicated by the IOC's insistence that the Games are free of politics (for example, Rule 51, section 3 of the IOC's Olympic Charter compels host cities to prevent protests of any kind near Olympic cities) – an initiative that may render any political statement a simultaneous act of Olympic resistance. For the purposes of the chart below, protests that may have drawn the attention of security planners – such as those surrounding the aboriginal Lubicon land claims during Calgary, 1988, protests over socio-economic inequality and the diversion of public monies to the Salt Lake City Olympics in 2002, or the 'No Olympics on Stolen Land' campaign at Vancouver, 2010 – are excluded. Instead, the focus is on actors and groups whose primary objective is an act of violence or sabotage *in the first instance*, rather than more legitimate protest movements that may become vehicles for more physical confrontations.

So what constitutes an 'Olympic-related' threat? In many ways this is an impossible question to answer. Complicating factors include events that have taken place in a host nation prior to a Games that have no apparent connection to the event yet still occupy their related security planning and threat assessments. The impact of attacks on two Air India flights, both originating from Canada on 23 June 1985, by extremist Sikh nationalists operating from British Colombia on security planning for the 1988 Calgary Games is an example of this dynamic. Another definitional difficulty relates to the proclaimed goals of terrorist actors. Some groups, such as the Catalan nationalists Terra Lliure or Greek leftists Revolutionary Struggle articulated clear hostility to the Olympics, whilst others were less explicit, yet had exerted a similar impact on security planning.

Finally, given the importance of projecting carefully managed brand images of host cities to their global audiences, Olympic organisers have consistently downplayed any connection between the Games and those that may resist them. For example, as competitors, spectators and officials filed through Tokyo's

Table 13.1

Olympiad	Perceived/actual threat and/or prior activity	Terrorist/other politically motivated attack
Munich, 1972	• Little perceived threat. 'Low-intensity' security model.	• Black September Organisation attack, kidnap and murder of 11 Israeli team members. • Violence by Irish protestors during men's cycling road race
Innsbruck, 1976	• Popular Front for the Liberation of Palestine (PFLP) hostage siege, Vienna six weeks before the opening ceremony. • Three BSO operations in Austria during 1973. • Potential Red Army Faction threat.	• None, but injuries through poor stadia design and crowd control.
Montreal, 1976	• Little (lapsed FLQ threat).	• None, but two security breaches.
Lake Placid, 1980	• Three Puerto Rican separatist (FALN) attacks, Chicago. • Two right-wing extremist (Omega-7) bombings, New York. All in the month preceding the opening ceremony.	
Moscow, 1980	• Risks associated with prior Soviet invasion of Afghanistan.	
Sarajevo, 1984	• Threat of ethno-nationalism and national fragmentation following the death of Tito. • Threats by Croat extremists.	
Los Angeles, 1984	• American Secret Army for the Liberation of Armenia (ASALA) makes explicit threats to Turkish athletes. • Remaining FALN threats. • Cold War tensions. • US involvement in Lebanon conflict.	
Calgary, 1988	• Prior Sikh nationalist activity operating from British Colombia.	

Seoul, 1988	• Public North Korean hostility.
	• Threats to aviation.
		• State terrorism (from North Korea) and proxy Japanese Red Army attacks:
		  i Bombing Korean Airlines KAL 858 at the cost of 115 lives, November 1987.
		  ii Disruptive global campaign against Seoul-bound airlines during 1988.

Albertville, 1992	• Environmentalist threats. • PFLP activity associated with threats to arrest leader George Habash in France. • 21 Iparretarrak ('Northern Basques') explosive operations in the 12 months leading up to the opening ceremony. • Corsican National Liberation Front (FLNC) operations in the lead-up to the Games.	• Coordination, Offensive, Use, Interruptions, and Cut (COUIC) sabotage of opening ceremony broadcast.

Table 13.1 continued

Olympiad	Perceived/actual threat and/or prior activity	Terrorist/other politically motivated attack
Barcelona, 1992	• Euzkadi Ta Askatasuna (ETA) bombing campaign (ETA also attacked Madrid's 2012 bid). • Escalation of Catalan nationalist (Terra Lliure) attacks 1987–1992. • Five bombings and one small-arms attack by Grupo de Resistencia Antifascista Primo Octobre (GRAPO) during the first half of 1992.	• ETA attack on electricity supply to the opening ceremony. • GRAPO double bombing of Catalan oil pipeline the day before the opening ceremony.
Lillehammer, 1994	• Site of Munich-related Mossad murder. • Risks associated with the signing of Oslo Accords, 1993.	
Atlanta, 1996	• 1993 World Trade Center Bombing (Violent Jihadi Extremist (VJE)). • 1995 Oklahoma Bomb (right-wing extremist (RWE)). • Khobar Towers attack (VJE). • Risks associated with chlorine transportation under the Olympic press centre. • Soldier guarding the Olympic village shot during Independence Day celebrations. • Security agencies' preoccupation with local gang activity.	• Eric Rudolph (RWE) bomb in protest against US abortion laws. • Armed man entering the opening ceremony. • Escalation of bomb threats after the Centennial Park bomb (suspicious packages reported every 10 minutes until the end of the Games).
Nagano, 1998	• Aum Shinrikyo attacks in Nagano prefecture during 1995.	• Left-wing rocket attack on Narita airport as athletes arrived four days prior to the opening ceremony. • Explicit (hoax) bomb threats to the Games.
Sydney, 2000	• Arrest of a right-wing extremist, charged with intent to cause explosions around the Olympic site.	• Alleged violent jihadi extremist attack averted.
Salt Lake City, 2002	• 9/11. • New York and Florida anthrax attacks, 2001.	• Anthrax hoax at Salt Lake City airport four days prior to the opening ceremony. • So-called 'beer riot'. • 600+ 'suspicious package' scares.
Athens, 2004	• 9/11. • Threat from VJE. • Potential response to sentencing of 17 November 17 activists, December 2003. • Five bomb attacks by leftist anti-Olympic 'Revolutionary Struggle' group in the 11 months leading to the start of the Games.	• Security breached during the marathon. • Hoax bomb threat during opening ceremony. • 82 bomb threats.

continued

Table 13.1 continued

Olympiad	Perceived/actual threat and/or prior activity	Terrorist/other politically motivated attack
Turin, 2006	• Italian Anarchist Federation (FAI) letter-bomb campaign against Turin's police, 2005. • Merging of environmental protest with Olympics. • Attempted VJE attacks on Milan's mass transit system, 2004. • Impact of 7 July and 21 July London attacks on security planning. • Iraq War (coalition member).	• FAI Anarchist attacks on Olympic store.
Beijing, 2008	• Chinese government identify Uyghur militants as principal threat to the Games.	• Uyghur militant improvised explosive device (IED) attack in Kashgar, killing 16 people, four days before the Games. • Two further attacks in Kuqa during the first four days of the Games.
Vancouver, 2010	• No specific threat (although high levels of public fear). • Potential opposition to military deployment in Kandahar as part of the NATO-led coalition.	
London, 2012	• Violent jihadi extremism. • Resurgent right-wing extremism. • Remaining violent dissident Republicanism.	
Sochi, 2014	• Geographical proximity to Chechen, South Ossetia and Abkhazian disputes.	
Rio de Janeiro, 2016	• Local crime. • Local criminal use of terrorist tactics including IEDs and 'man-portable air defence systems' (MANPADs).	

Source: Fussey *et al.* (2010).

Narita airport five days before the opening ceremony of the Nagano Winter Olympics, 1998, left-wing extremists fired three home-made rockets into the complex's cargo area. Whilst Japanese authorities were quick to highlight that the perpetrators, the 'Revolutionary Workers Association', were protesting against Narita's expansion and therefore not committing an Olympic-related attack (and tarnishing the Games accordingly), its temporal proximity to the event guaranteed a level of global exposure that would have been difficult to otherwise achieve. In sum, in an attempt to avoid becoming mired in definitional debates over the nature of terrorism (which are ably addressed elsewhere, for example, Silke, 2004; Hoffman, 2006) and what constitutes an 'Olympic-related' threat, what follows is an outline of resistance activity that gives primacy to the application of violence as first resort and was of significant concern to Olympic security planners.

Aside from the prominent and much-discussed attacks at Munich and Atlanta, *almost all* terrorist activity surrounding the Olympics has occurred outside the time and place of the Games, thus illustrating the importance of the *spatial* and *temporal* displacement of attacks. Amongst these, it is argued that the Seoul and Barcelona Olympiads have probably experienced the most significant intensity of displaced terrorist violence. Also notable is the diversity of groups who have used the Olympics to project their aims more broadly and increase the potency of their threat. Despite the propensity of different terrorist ideologies to influence a varied selection of targets (inter alia Drake, 1998; Fussey, 2010), the Olympics provide a ready and consistent symbolic target for a variety of groups, regardless of their ideological, operational and tactical diversity, as illustrated above. Finally, in taking a broad view of terrorist threats to the Olympics over a long period, it can be argued that, with notable exceptions, many of the groups that target them originate from the local socio-political contexts of the host nations. This marks an interesting contrast to the conspicuous internationalism of the Olympic Games. Moreover, for many, such as Euzkadi Ta Askatasuna (ETA) (targeting Barcelona, 1992, and Madrid's bid to host the 2012 Games) and Eric Rudolph (Atlanta, 1996), the Olympics was only one episode that punctuated a much longer campaign. These threats are presented as follows:

References

Drake, C. (1998) *Terrorists' Target Selection*. Basingstoke: Palgrave-MacMillan.

Fussey, P. (2010, in press) 'An Economy of Choice? Terrorist Decision-Making and Criminological Rational Choice Theories Reconsidered', *Security Journal* 23.

Fussey, P., Coaffee, J., Armstrong, G. and Hobbs, R. (forthcoming 2010) *Sustaining and Securing the Olympic City: Reconfiguring London for 2012 and Beyond*. Aldershot: Ashgate.

Hoffman, B. (2006) *Inside Terrorism*. New York: Columbia.

Silke, A. (2004) 'Terrorism and the Blind Men's Elephant.' In Alan O'Day (ed.) *Dimensions of Terrorism*. Aldershot: Ashgate, pp. 241–257.

Index

Page numbers in *italics* denote tables, those in **bold** denote figures.

9/11 39, 143; lessons of Olympics since 42–3
Abdul Mutallab, Umar Farouk 82
accidents 146–7
Afghanistan, British troops in 24–5
Agamben, G. 119
Agnew, Jonathan 22
Ahmed, M. 109
airline ticketing/profiling procedures 82–3
airport security 64, 78–9, 81–2
Airwave/AirwaveSpeak radio systems 189–91
al Zawahiri, Ayman 40, 53, 56, 59
Al Qaeda Associates 40–1
Al Qaeda Core 40, 41–2, 43
Al Qaeda in the Islamic Maghreb 40–1
Al Qaeda: aims of 37–8; attacks 5–6; challenges posed by 40–2; decision to act 43–4; ideological inheritance 34–6; and London Olympics (2012) 23–6, 42–3; prisoners associated with 229; target selection 53–4, 55–7, 63; terrorist strategy 38–40; tunnel-vision effect 229
Al Qaeda-inspired groups and individuals 25, 41–2
Al-Qurashi, Abu 'Ubeid 56–7
Albertville Olympics (1992) 29, *240*
Alison, Laurence 189, 191
Alkibiades 17–18
Allison, Chris 180, 197
Altshuler, Alan A. 137, 147
American Civil Liberties Union 100
Amsterdam Treaty 208, 213, 215–16
Ancient Games 17–18, 19–20, 180
Anti-Doping Convention, EU 155
Arbuthnott, Kevin 193

Armstrong, G. 93, 94, 97, 98
'Army of God' 4
Arup 145, 155–6
Ashraf, Afzal 7–8, 24, 32–46
Association of Chief Police Officers (ACPO), 187, 194–5, 196, 199; 'Secure by Design' scheme 120, 125, 128–9; Olympic Preparation Committee 198
Aston, C. 6
Athens Olympics (2004) 42–3, 76, 77, 95, 101–2, 120, 122–3, 150, 152, 154, 208, 216–17, 218, *241*
Atkinson, M. 7
Atlanta Olympics (1996) 3–4, 18, 57–8, 60, 61, 108, 122, 137, *241*
ATOS Origin 155, 177
audit, rise of 152–3
Aum Shinrikyo 4, 24
Automatic Number Plate Recognition (APNR) 94, 126, 127
Averoff, M. George 149
aviation transportation systems 78–9, 81–3, 88

Baash, S. 110, 234
Bache, I. 210
Ball, K. 95, 99
Barcelona Olympics (1992) 29, 122, 151, *241*
Barney, Robert K. 20
barriers, use of 67–8, 77
Barry, John M. 142
Bartlett, D. 102
Baskin, D. 52
Beard, Matthew 156
Beck, Ulrich 95, 135
Beckett, A. 125

Beijing Olympics (2008) 42–3, 59, 76, 103–4, 137, 151, 152, 154, 155, *242*
Bellavita, C. 6
Bennetto, J. 98
Berlin Olympics (1936) 18, 149
Bernstein, Peter L. 135
Bin Laden, Osama 28, 36–8, 40
Black September Organisation 34, 96, 103
Body-Gendrot, S. 209
Bosher, L. 121
boycotts 20, 137, 143, 153
Boyle, P. 7, 118, 123
Bradsher, K. 104
Brasch, R. 17
British Olympic Authority (BOA) 186
British Security Industry Association 170
British Transport Police 79
Brodeur, J.-P. 221
Brown, Gordon 24, 25, 121, 122
Browning, M. 119
Buck, Graham 152
Buntin, J. 6, 108, 233
Buzan, Barry 148

'C41' programme 101–2
Calgary Olympics (1988) *240*
California Olympiad Bond Act (1927) 149
Carter, H. 92
Cauley, J. 64
'CCTV Challenge Scheme' 94
Centre for the Protection of the National Infrastructure (CPNI) 77
Chakravarthi, R. 76
Charter Amendment 'N', Los Angeles 150
Charters, D. 6
chemical detection systems 80
Chesterman, S. 221
Civil Contingencies Act (2004) 199
Civil Protection Mechanism, EU 217
Clarke, R. 50, 51–2, 62–3, 96
CLM consortium 155–6, 166
close protection 172–3
closed-circuit television (CCTV) 92, 93–5, 96–104, 105–7, 108–9, 123–4, 125
Coaffee, Jon 7, 8, 95, 101, 102, 118–29
Cold War 18
Collins, Eamon 55–6
command and control structure 172, 191–4
Command, Control, Plan and Resource Programme 184
communication 189–91
Communities and Local Government (CLG) Emergency Room 196

community engagement, London Olympics 44–5
Concept of Operations (CONOPS) 167
CONTEST strategy 228–9
contracting out 155–6
control and risk 135–6
Coogan, T.P. 55
Cook, I. 103, 104
Coordinating National Research Programmes on Security during Major Events in Europe manual 211
Copeland, David 106, 108–9
Cornish, D. 50, 51–2, 62
Corporate Manslaughter and Corporate Homicide Act (2007) 189
Cottrell, R.C. 7
Counter-Terrorism and Extremism Liaison Officers (CTELOs) 213–14
counter-terrorism, target hardening 63–8
Countering International Terrorism: the UK's Strategy (2006) 228
Crawford, A. 208
Crego, Jonathan 189, 191
Crenshaw, Martha 50
Crime Prevention through Environmental Design (CPTED) 120, 128–9
crowded places: defending Olympics 122–4; East London 124–8; main stadium and other Olympic venues 124; paradox of defending 120–2
Culf, A. 124, 127
Cultural Olympiad 174, 177
cyber security 172

data overload, CCTV 108–9
Davis, M. 94
de Coubertin, Pierre 20, 136, 149
de Menezes, Jean Charles 181, 200
De Waard 221
Decker, S. 6, 105, 232
Deobandi Taliban 35–6
Department for Communities and Local Government (DCLG) 196
Department for Culture, Media and Sport (DCMS) 186, 198
Department for Transport 79
DiMaggio, Paul J. 148, 157
diplomacy 18
displacement effect 21, 64–8, 107–8, 127, 233
dissident Irish Republicans 5, 26–8, 228, 229, 230
Ditton, J. 94
Dolnik, A. 76

Donahue, Amy 187, 191, 192, 200
doping 154–5
Dorn, N. 221, 222
Doyle, Kate 154
Drake, C.J.M. 59, 60–1, 84, 243

East London: organised crime 232; securing 124–8; surveillance experiment 94
economic evaluation 151–2
Edwards, A. 208
Eick, V. 110
ekecheiria 19–20, 180
Eleftherotypia 232
embassy security 64–5
emergency services 191–6
employment 234–5
Enders, Walter 50, 64, 65, 69
Engage Programme 184–5
entrepreneurialism 149–51
Espy, Richard 143
ETA 5, 21, 51
ethics and surveillance 104, 109–11
'Euro 96' football championships 98
European National Football Information Point 212–13
European Union (EU): event security partnerships 218–19; information/research sharing networks 213–15; international football games 212–13; private security industry as partner 219–22; terrorist attacks 5
European Security Research and Information Forum (ESRIF) 222
European Security Strategy (2003) 208–9
Europol 211, 212–13, 214, 216
Evans, David 9, 163–79
Evans, G. 103
Evans, J. 6
event size 139–40
exercises 198–9
external–national risk 140–2
external–transnational risk 143–4

Facial Recognition CCTV (FRCCTV) 95, 99–101, 106–7
Farrington, D.P. 106
feasibility studies 145, 146
Feeley, M. 110
FIFA World Cup (2006) 98, 110, 111, 212
Fire and Rescue Service: National Coordination Centre 196; New Dimension Programme 195–6
Fire and Rescue Services Act (2004) 196

Fire Services Act (1947) 195
Fisher, A. 221
Flyvbjerg, Bent 137, 144, 147
football policing network, EU 212–13
Football Policing Unit 212
Football Trust 97
football-related CCTV surveillance 97–9
Foreign Relations Authorization Act (1990), US 104
Foucault, M. 92
Franco-British Exhibition 149
Front for the Liberation of the Enclave of Cabinda (FLEC) 22–3
Fussey, Pete 1–10, 91–111, 118, 119, 127, 128, 165, 175, 227–36, 239–43
Fyfe, N.R. 93

Gamarra, A. 7
Gates, K. 107
Generic Regional Response Plan (GRRP) 199
George, S. 210
Geyer, F. 208
Ghaffur, Tarique 125, 127
Giddens, Anthony 135
Gill, M. 106
Gilmour, R. 49
Giulianotti, R. 7, 97
government departments, role of 167–8
Government Olympic Executive 183
government: role of 182–9; targeting 58
Graham, S. 93
'Grand Beijing Safeguard Sphere' 103
Graycar, A. 232
Gregory, Frank 9, 208–23
Grieve, John 185–6
Groussard, Serge 151
Gruszczak, A. 209
Guild, E. 208
Guttmann, Allen 137, 138

Habash, George 20–1
Hacking, Ian 135, 136
Hagemann, A. 98, 110, 234
Haggerty, K. 7, 118, 123
Hakala, J. 220, 221
Hamas 36
Hancox, P.D. 97
Handbook for Member States Co-operation Against Terrorist Acts (EU) 210, 215–17, 218
Harris, N. 59
Hayes, B. 222
Hazard Assessment 185

Health and Safety Executive (HSE) 182
Hess, K.M. 219
Hezbollah 36
Higgins, David 137
Higgs, Robert 147
hijackings 84
Hill, Christopher R. 18–19, 20, 143
Hillsborough tragedy (1989) 98–9
Hilton, Christopher 18
Hinds, A. 7, 95, 102, 103
HM Inspectorate of Constabulary (HMIC) 181–2, 200–2
Hobbs, D. 94
Hoffman, Bruce 5, 21, 25, 107, 242
Home Office 98, 99, 122, 128, 199; Olympics Safety and Security Strategy Group 198; responsibility for security 186–7; Scientific Development Branch (HOSDB) 77
Home Secretary 182
Hood, Christopher 136, 147–8
Horne, C. 93
Hughes, G. 208
human adaptation to technological innovation 107–8

Ibrahim, Andrew 41, 42, 44, 45
Ibrahim, Muktar Said 86
Identify and Disrupt Programme 184
identity cards 79
ideological inheritance, Al Qaeda 34–6
ideologues, role of 45
ideology: and target selection 52–63; willingness to buy into 228
Im, E.I. 64
Impact Assessment 185
Indian Commonwealth games (2010) 22
industry capacity 175–6
information/research sharing networks 213–15
Innsbruck Olympics (1976) *240*
insecurity, shifting landscapes of 105–6
insurance 152
insurgency-based warfare 39
Intelligence and Security Committee 85
intelligence exchange agreements 154
inter-agency coordination: appendix 202–4; communication 189–91; interoperability 200–2; leadership 191–4; logistics and resource management 194–6; planning 196–8; public relations 199–200; role of government 182–9; training and exercising 198–9

Inter-Agency Liaison Officer (ILO) scheme 195
internal–national risk 139–40
internal–transnational risk 142–3
International Olympic Committee (IOC): approach to risk 146; *Candidature Procedure and Questionnaire* 153; Charter 20, 104, 111, 239; Coordination Commission 153; dependence on revenues 142, 147, 152; Evaluation Commission 153; responsibility for security policy 164–5
International Treaty Against Doping in Sport (2005) 155
Internet, role of 20–1, 45
interoperability 181, 200–2
Introna, L.D. 100–1
Iraq, invasion of 24–5, 39
Irish Republican Army (IRA) 21, 26–8, 55–6, 65–6; *Green Book* 54–5; *see also* dissident Irish Republicans; Provisional IRA
Islamism 32–3, 34–6
'island security' 123
Israel: construction of barrier 67–8; suicide attacks against 63
Iton, J. 151

Jackson, R. 16
Jacobs, S. 50, 52
Jemaah Islamaya 21
Jenkins, Brian 24
Jennings, Will 8–9, 119, 135–57, 211, 217
jihad 5–6, 33, 34–6, 38
Johnston, L. 102, 118, 122, 123
Joint Terrorism Analysis Centre (JTAC) 6, 184, 212
Jones, T. 136, 219, 221
Juska, A. 220
just retribution 37

Kasperon, Roger E. 135
Keeler, Simon 41
Khan, Mohammed Sadiq 39, 85–6
Klausner, F. 7, 98, 105, 110
Klinenberg, Eric 140
Kuming city attack (2008) 43

Lagadec, Patrick 135, 140
Lahore attack (2009) 21, 22, 28, 43
Lake Placid Winter Olympics (1980) 97, *240*
Lammer, M. 19
leadership 191–4

Index

Lee, C. 110, 234
Lenckus, Dave 152
Lend Lease 151–2
Lenskyj, H. 104, 110, 234
lessons learned 187–9
Levi, M. 221, 222
Lillehammer Olympics (1994) *241*
Lindsay, Jermaine 41, 85–6
Lisbon Treaty (2009) 208, 209
Lockerbie bombing (1988) 75, 82
Lodge, Martin 119, 138, 144, 211, 217
logistics 194–6
London 2012 Olympic and Paralympic Safety and Security Strategy (2009) 128, 217–18, 229–30, 223
London 2012 Organising Committee (LOCOG): dependence on revenues 142; risk-management 153, 186; role of 217; safety and security responsibilities 183
London bombings: (1992/93) 65–6; (1996) 66–7; (2005) 6, 39, 75, 78, 85–6, 87, 119, 121, 189, 229
London Fire Brigade (LFB) 195
London Host City Contract 138, 180–1
London Olympics (1908) 149
London Olympics (2012): and Al Qaeda 23–6, 42–5; critical reflections on private security 174–8; key security partners 164–5; national and local contexts for surveillance 93–5; private security requirements during build 171; private security games requirements 172–3; risk transfer 151–2; securing East London 124–8; security legacies 128–9, 234–5; symbolic significance as terrorist target 6
London Regional Resilience Forum (LRRF) 199
London Strategic Emergency Plan 199
London underground system 78
Los Angeles Olympics (1932/1984) 149, 150, *240*
Loucopoulos, P. 76
Luberoff, David 137, 147
Luckes, David 144, 145, 146
Lyon, D. 100, 107, 110

McCahill, M. 93, 98, 104
MacAloon, John J. 149
Macartney, J. 103
McGladdery, G. 56, 65, 230
Macko, S. 122
Macpherson, W. 110, 234

Madrid bombing (2004) 41
Magnay, J. 137
Magnier, M. 103
Majone, Giandomenico 136, 147–8
major-event (MES) security, EU 208–23
major incidents, responses to 192–4
marine transportation systems 83–4
market, transfers to risk to 151–2
Marsh, P. 98
Martin, Simon 18
Marx, G.T. 110
mass casualty attacks 28, 38–9, 40, 53, 85
Mathieson, T. 92
media publicity 20–1, 24, 27–8, 34, 55, 56–7, 58–9, 242
metal detectors 64
Metropolitan Police 79, 109, 180; Anti-Terrorism Branch (SO13) 109; Counter-Terrorism Command (SO15) 213, 214; National Olympic Oversight Group 198; reviews of 181–2, 200–2
Metropolitan Police Authority (MPA) 181, 197, 198
Mexico City Olympics (1968) 154
MI5 6, 24, 228
Michel-Kerjan, Erwann O. 135
Migdalovitz, C. 216, 217
Military studies in the jihad 54
military weapon, sport as 17–19
Miller, Stephen G. 17–18, 20
Ministerial Committee on Economic Development Sub-Committee on the Olympic and Paralympic Games (ED(OPG)) 182
Ministerial Committee on National Security, International Relations and Development Sub-Committee on Protective Security and Resilience (NSID(PSR)) 182
Ministry of Defence 197–8
Minnaar, A. 124
mobile surveillance vehicles 98
modernity and risk 135–6
Monar, J. 208
Montreal Olympics (1976) 96–7, 150, *240*
Moran, Michael 136, 144, 148
Morgan, J.B. 97
Moscow bombings (2004) 79, 87
Moscow Olympics (1980) 18, 152, *240*
multi-agency working, problems affecting 191
multi-event thwarting 64
multi-wave attacks 77
Mumbai attacks (2006/2008) 22, 28, 76, 84

Munich Olympics (1972) 1–3, 4, 20, 21, 27–8, 33–4, 56–7, 60, 108, 137, 138, 150–1, 235–6, *240*
Muslim communities 44–5
Muslim world, driving Western forces from 37–40

Nagano Olympics (1998) *241*, 242
nation-building 18
National Audit Office 143, 145, 153
National Coordinator for Domestic Extremism (NCDE) 214
National Counter Terrorism Security Office (NaCTSO) 121; *Counter Terrorism Protective Security Advice for Stadia and Arenas* 125–6
National Olympic Coordination Centre 195
National Olympic Security Coordinator 197
national points of contact (MFIPs), EU 212–13
National Police Improvement Agency (NPIA) 189–90
National Public Order Intelligence Unit (NPOIU) 214
National Reporting Centre (NRC) 195
National Research Council 80
Nemean Games 19–20
Newburn, T. 219, 221
Newman, G. 52, 62–3
non-games venues 173–4
non-state terrorism 16
Norris, C. 93, 94

O'Hare, P. 122
O'Malley, Pat 136
Ocqueteau, F. 209
Office for Security and Counter Terrorism (OSCT) 182–3, 187; Industry Advisory Group 170; Olympic and Paralympic Security Directorate (OSD) 182–3, 187–8, 196, 197, 198–9, 229–31
Olympic Advisory Group 102
Olympic and Paralympic Policing Coordination Team (OPC) 187, 196, 197, 198–9
Olympic and Paralympic Safety and Security Strategy 182–9; Senior Responsible Owner (SRO) 182
Olympic Board 186, 198
Olympic Delivery Authority (ODA) 198; contracting 105, 151; role of 166–7; safety and security responsibilities 171, 183

Olympic Games organising committee (OCOG): financial risk 150; and security 165, 170, 171, 172
Olympic Intelligence Centres (OICs) 102, 154, 185
Olympic Programme Support Unit 198
'Olympic rings of security' 233
Olympic Safety and Security Strategic Risk Assessment (OSSSRA) 185–6, 198–9, 230; work stream structure *231*
Olympic sites, resilient design of 118–29
Olympics: Ancient Games and *ekecheiria* 19–20; bidding process 123–4, 144–5, 152–3; defending crowded spaces of 122–4; defending main stadium and other venues 124; EU and security 215–18; features of terrorism and their implications for 16–17; organisational responses to risk 147–56; reflections on terrorism and security 230–3; and risk 63, 136–8; security studies 6–7; surveillance prior to 2001 96–9; surveillance since 2001 99–104, 105–11; targeting 58–60, 61; terrorist threats (1972–2016) 239–43; towards an understanding of terrorism and 1–10; two dimensions of risk 138–44
'Olympism' fundamental principles of 19, 20
Omagh bombing (1998) 26–7
Operation Harvest 98
Operation Wrath of God 2–3
optimism bias 144–5
organisational growth **141**
organisational responses to risk 147–56
organised crime 232
organisations 202–4
'Oyster' system 78, 80

Pakistan, terrorist links to 25
Palestine Liberation Organisation (PLO) 3, 34, 36
Palestine, suicide attacks on Israel 67–8
Pape, R. 52, 85
Paralympics 59
passenger screening 79–80, 81–2, 83, 84
Payne, Michael 136, 146
peaceful mission, Olympic Games as 19–20
Peek, L. 120
perimeter security 97, 123, 127, 172
Perrow, Charles 135, 144, 146
person-borne improvised explosive devices (PBIEDS) 120

philanthropy 149–51
Pidgeon, N.F. 135
planning: security 196–8; terrorist attacks 59–61
Police National Information and Coordination Centre (PNICC) 194–5
Police Working Group on Terrorism (PWGT) 213
police: coordination of 194–5; resources of 176; role of 167; *see also* Metropolitan Police
Policing critical incidents 185–6
policing network, football 212–13
political 'status quo' 33–4
political terrorism 36–7
political weapon, sport as 17–19
Pollitt, Christopher 147
Pope, S.W. 18
Popular Front for the Liberation of Palestine 20–1
Porter, Theodore M. 135
Pound, Richard W. 135, 138, 145
Powell, Walter W. 148, 157
Power, Michael 136, 148
Prekas, N. 76
Prepare Programme 184
Preuss, Holger 140, 151
prime contractors 168–9
Private Military and Security Partnerships (PMSCs) 220
private security industry (PSI): composition 168; contracting out 155–6; existing customers and 'business as usual' 176; games requirements 172–3; London 2012 and beyond 178–9; as major-event security partner 219–22; non-games requirements 174; and planning 177–8; previous experience at major events 174–5; prime contractors and SMEs 168–9; requirements during the build 171; security industry regulation 171; sponsor effect 169; trade associations 170; understanding industry capacity 175–6
private-sector cohesion 178
procurement: problems 175–6; process 170; and sponsors 177
Proposals Relating to the Enhancement of Measures to Counter Football Violence, EU 210
Protect Programme 184
protest groups 213
Provisional IRA (PIRA) 55–6, 94, 106, 107–8, 120

psychological aspects of suicide terrorism 86–8
public order, information/research sharing networks 213–15
public relations 199–200
public sector, integration with 178
public security players 164–8
public-private partnerships (PPPs) 218–22
Pursue Prevent Protect Prepare, the UK's strategy for Countering International Terrorism (2009) 228

Quran 35, 38

radio systems 189–91
rail transportation systems 78–81
rational choice approach 34, 50–2
Rawlinson, P. 232
Real IRA 26–8
Reeve, Simon 2, 3, 6, 150
regulation: rise of 152–3; security industry 171; sport 154–5; transnational networks of 153–6
Reid, Richard 41, 81
Reilly, Nicky 41, 42, 45
Reiner, R. 110
religious terrorism 4–6, 36–7
Resilience Suppliers' Community (RISC) 170
resilient design 118–29
Richards, Anthony 1–10, 15–29, 227–36
Rigg, Nancy J. 156
'Ring of Steel', London 65–6, 67, 120
Rio de Janeiro Olympics (2016) 128, *242*
risk aversion 146
risk discounting 145
risk: disparate risks 230–2; and modernity/control 135–6; normal accidents 146–7; and Olympics 136–8; optimism bias 144–5; organisational responses at Olympics 147–56; temporal and spatial 232–3; two dimensions of 138–44
risk-assessment 185–6
risk-management: rise of 152–3; transnational networks of 153–6
Riyadh, terrorist attacks (2003) 77
road transport system 76–8
Roberts, Dexter 156
Rogers, P. 123
Roy, Olivier 41
Rudolph, Eric 4, 57–8, 61, 108

sabotage 28–9
Sack, Kevin 146

Sadlier, D. 6
'Safe Cities' programme, China 104
safe legacy, ensuring 128–9
Safer Places; the Planning System and Crime Prevention guide 122
Sageman, Marc 25, 42
Salafism 33, 35–6
Salt Lake City Olympics (2002) 100, 105, 146, 150, 156, *241*
Samatas, M. 101, 102, 110, 119, 123
Sanan, G. 7, 18, 28, 29, 95, 108
Sandler, Todd 50, 65, 69
Sarajevo Olympics (1984) *240*
Saudi Arabia, Western forces in 37–8
Scarman, L. 110, 234
Schreiber, Manfred 151
Scott, James C. 135, 144
Seabrook, T. 94
securitisation and risk responses 156
security budgets 156
security directorate 167–8
security guards 175
Security Handbook for the Use of Police Authorities and Services at International Events (EU) 209–10
security legacies, London Olympics 110, 127, 128–9, 234–5
security partners/partnerships 164–5, 177–8
security procedures, changing routine of 87–8
security training, EU 220–1
security: holistic approach to 177–8; shifting landscapes of 105–6; supplying 106–8
Seoul Olympics (1988) *240*
sex industry 232
sharia 34–6, 38
Sherman, J. 109
Shute, S. 92
Silke, Andrew 1–10, 21, 49–70, 81, 86–7, 227–36, 242
Simon, J. 110
simultaneous attacks 28
single-event thwarting 64
Sinn, Ulrich 19
situational crime prevention 51–2, 62–3, 65–8
small and medium-sized enterprises (SMEs) 168–9
Smith, G.J.D. 92
Sochi Olympics (2014) *242*
Sommers, I. 52
South Africa, suspension from Olympics 19

spatial threats 232–3
spectator searches 165, 172, 175
Spivey, Nigel 17
sponsor effect 169
sport: as political and military weapon 17–19; regulation of 154–5; surveillance prior to 2001 96–9; surveillance since 2001 99–104; targeting of 20–2, 55–8; targeting of as developing trend 22–3
Spriggs, A. 106
stadia: defending 124; design/construction 125–6, 166–7
standardised security strategies 105–6
state sponsors 149–51
state-led organisation 149–51
Steele, J. 102
stewards 165, 175
Stockholm Programme 208–9
Strategic Threat Assessment 185
Strategy for Countering International Terrorism 228
Sturcke, James 22
suicide terrorism 52, 53, 63, 67–8; and transport system 84–8
Sullivan, J.E. 149
Summer Olympics 58–9
Superbowl XXXV (2001) 99–101
surveillance technology and its application 92–3
surveillance: in context 92–5; and ethics 109–11; and major sporting event security after 2001 99–104; and the Olympic spectacle 95–9, 105–11; scale and reach of 108–9
Suskind, Ron 146
Swain, Steve 8, 75–89
Sydney Olympics (2000) 95, 122, 137, 150, 152, *241*

tafkir 36
Tailby, R. 232
target reconnaissance 85–8
target selection : calculus of 49–51; focusing on environment 51–2; and ideology 52–63; London Olympics 20–2; target hardening 63–8
Taylor, Max 50, 52
Taylor, Peter 21
Taylor, T. 7
technological failures 102
technological surveillance 92–111
temporal threats 232–3
Terrorism Act (2000) 76

terrorism: features and implications for Olympics 16–17; information-/research-sharing networks 213–15; justification for 36–7; and the Olympics 1–10, 33–4; reflections on 230–3
terrorist attacks, limits on 55–6, 59
terrorist groups, capability of 60–1
terrorist manuals 54–5, 59–60
terrorist operation, stages of 59–61
terrorist tactics 28–9; Al Qaeda 38–40
terrorist threats (1972–2016) 239–43
Thompson, A. 6, 7, 118
threat level 184
Tilley, N. 106
Togo national football team 22–3
Toohey, K. 7
Töpfer, E. 110
trade associations, role of 170
training 198–9
transport security 173
transport system 75–89; and suicide terrorism 84–8
Tuohy, Robert 187, 191, 192, 200
Turin Olympics (2006) 212, 218, *242*
Turkish Islamic Party 43

UEFA European football championships (2008) 110
UK Select Committee on Culture, Media and Sport 150
UK, technological surveillance 93–5
United Nations Interregional Crime and Justice Research Centre (UNRICI) 211
US Science Applications International Corporation (SAIC) 101–2

Vancouver Olympics (2010) 77, 153, *242*
Vaughan, Diane 135
vehicle searches 77–8

vehicle-borne improvised explosive devices (VBIEDS) 118, 120, 125–6
VIPs 169, 172–3
Vlachou, E. 7, 95, 102, 103
Vulnerability Assessment 185

Wahabism 33
Wakefield, A. 221
Waldheim, Kurt 17
Walker, C. 76
Walker, N. 208
Walsh, D. 51
War on Terror 39, 40
Wardlow, G. 17
Wattis, L. 94
Webster, C.W.R. 93
Webster, F. 95, 99
Wells, A. 109
Welsh, B.C. 106
West, Lord 118, 186
West, attacks in 39–40, 41
Weston Keith 9, 180–204, 229
Williams, C. 93
Wilson, J. 94
Winter Olympics 58–9
Wolff, A. 1, 2, 60, 235
Wood, D. 100–1, 118, 123
World Anti-Doping Agency (WADA) 154–5
Worthy, W. 55

Young, David C. 137, 149
Young, K. 7
Yu, Y. 6

Zelukin, Michael 77
Zeus 19
Zionism 36

eBooks – at www.eBookstore.tandf.co.uk

A library at your fingertips!

eBooks are electronic versions of printed books. You can store them on your PC/laptop or browse them online.

They have advantages for anyone needing rapid access to a wide variety of published, copyright information.

eBooks can help your research by enabling you to bookmark chapters, annotate text and use instant searches to find specific words or phrases. Several eBook files would fit on even a small laptop or PDA.

NEW: Save money by eSubscribing: cheap, online access to any eBook for as long as you need it.

Annual subscription packages

We now offer special low-cost bulk subscriptions to packages of eBooks in certain subject areas. These are available to libraries or to individuals.

For more information please contact webmaster.ebooks@tandf.co.uk

We're continually developing the eBook concept, so keep up to date by visiting the website.

www.eBookstore.tandf.co.uk

Printed in Great Britain
by Amazon